For a complete listing of the
*Artech House Power Engineering Library*,
turn to the back of this book.

# Introduction to Power Electronics

Paul H. Chappell

**ARTECH**
**HOUSE**

BOSTON | LONDON
artechhouse.com

**Library of Congress Cataloging-in-Publication Data**
A catalog record for this book is available from the U.S. Library of Congress.

**British Library Cataloguing in Publication Data**
A catalog record for this book is available from the British Library.

ISBN-13: 978-1-60807-719-9

**Cover Design by Vicki Kane**

© **2014 Artech House**

10 9 8 7 6 5 4 3 2 1

*To Roma, Jonathan and Kate, and Edward*

# Contents

**Preface**      **11**

**1**    **Introduction**      **13**

1.1    Power Semiconductor Devices      14
1.2    Power Conversion      14
1.3    Passive Components      16
1.4    Parameters and Analysis of Waveforms      17
1.5    Ideal Power Device      17
1.6    Practical Device      18
1.7    Sources of Information      19
     Selected Bibliography      19

**2**    **Diode**      **21**

2.1    Ideal Characteristic of a Diode      21
2.2    Simple Diode Model      21
2.3    Diode Model      23
2.4    Power Dissipation      25
2.5    Free-Wheeling Diode (Fly-Wheel)      26
2.6    Zener Diode      27
2.7    Schottky Diode      28
     Selected Bibliography      28

| | | |
|---|---|---|
| **3** | **Thyristor** | **29** |
| 3.1 | Gate Isolation Circuit | 31 |
| 3.2 | Snubber | 32 |
| 3.3 | Natural Commutation | 33 |
| 3.4 | Single-Phase AC Supply with Single-Gate Pulse Applied to a Thyristor and Resistive Load | 33 |
| 3.5 | Single-Phase Supply with Single-Gate Pulse Applied to a Thyristor and Inductance Load | 36 |
| 3.6 | Forced Commutation | 41 |
| 3.7 | Triac | 42 |
| 3.8 | Gate Turn-Off Thyristors | 42 |
| | Selected Bibliography | 44 |
| **4** | **Transistors** | **45** |
| 4.1 | Safe Operating Area | 45 |
| 4.2 | Snubber | 47 |
| 4.3 | Metal Oxide Semiconductor Field Effect Transistor | 52 |
| 4.4 | Insulated Gate Bipolar Transistor | 56 |
| 4.5 | Guide to Power Level and Frequency | 58 |
| | Selected Bibliography | 58 |
| **5** | **Heating and Cooling** | **59** |
| 5.1 | Thermal Resistance | 59 |
| 5.2 | Equivalent Circuit | 60 |
| 5.3 | Transient Thermal Resistance | 61 |
| | Selected Bibliography | 63 |
| **6** | **Phase-Controlled Thyristor Converters and Diode Rectifiers** | **65** |
| 6.1 | Phase-Controlled AC Thyristor Converter | 65 |
| 6.2 | Single-Phase Converter | 67 |
| 6.3 | Three-Phase Converter | 70 |
| 6.4 | Overlap | 77 |
| 6.5 | Inversion | 81 |

| | | | |
|---|---|---|---|
| | 6.6 | Power Systems | 82 |
| | 6.7 | DC Motor Drive | 84 |
| | | Selected Bibliography | 86 |
| **7** | | **Cycloconverter** | **87** |
| | 7.1 | Single-Phase Cycloconverter | 87 |
| | 7.2 | Three-Phase Cycloconverter | 91 |
| **8** | | **Inverter** | **93** |
| | 8.1 | Single-Phase Inverter | 93 |
| | 8.2 | Three-Phase Inverter | 93 |
| **9** | | **DC-to-DC Converters** | **99** |
| | 9.1 | Inductive Load | 99 |
| | 9.2 | DC Machine | 103 |
| | 9.3 | Regeneration | 110 |
| | 9.4 | Step-Up and Step-Down DC-to-DC Converters | 112 |
| | 9.5 | Step-Down (Buck) Converter | 112 |
| | 9.6 | Step-Up (Boost) Converter | 120 |
| | | Selected Bibliography | 128 |
| **10** | | **Systems and Methods** | **129** |
| | 10.1 | Logic Switching | 129 |
| | 10.2 | Defibrillator and Transcranial Magnetic Stimulator | 130 |
| | 10.3 | Synchronization to an AC Supply | 135 |
| | 10.4 | Measurement of Current and Voltage | 141 |
| | 10.5 | Lowpass Filter | 144 |
| | 10.6 | Circuit Protection | 145 |
| | 10.7 | Fourier Series | 147 |
| | 10.8 | AC Line Transient Suppression | 150 |
| | 10.9 | Efficiency of DC Converters | 150 |
| | 10.10 | Chaotic Behavior in Diode Circuits | 153 |
| | | Reference | 156 |
| | | Selected Bibliography | 156 |

**11      Examples                                        157**

Example 11.1                                       157
Example 11.2                                       158
Example 11.3                                       160
Example 11.4                                       160
Example 11.5                                       162
Example 11.6                                       164
Example 11.7                                       165
Example 11.8                                       166
Example 11.9                                       167
Example 11.10                                      169
Example 11.11                                      171
Example 11.12                                      177
Example 11.13                                      185
Example 11.14                                      188
Comment on the Fourier Series Examples             190

**List of Symbols                                     191**

**About the Author                                   195**

**Index                                              197**

# Preface

This book contains a description of the characteristics of different types of power semiconductor devices and their application to power converter circuits. It has been written as an introductory text containing the useful concepts and building blocks that go into making a power converter operate successfully. Device physics and manufacturing of semiconductors are deliberately not included, as these are detailed in other texts and are often studied separately. Some applications to power transmission, electric drives, and medical equipment are included to illustrate the wide range of power electronics in both small- and high-power circuits. Power electronics for large converters are housed in purpose built cubicles and buildings. Smaller converters are found in domestic white goods and in mobile devices. Large-scale power electronics and their optimization are an important link in the distribution of renewable energy to domestic and industrial environments. Smaller scale power electronics are becoming increasingly important in the automotive industry as vehicle electrification and hybridization become more widely adopted.

The design process for power electronics encompasses stages of conception, analysis, computer simulation, and experimentation. During this process, calculations and theoretical simulations should interact with practical testing. (The latter of which can be particularly exciting and enjoyable when the time comes to make that final connection of power to a new circuit; a moment often accompanied with some trepidation and a sharp intake of breath!) Practical testing is undertaken with a certain amount of caution because dangerous voltages and high currents (hot components) can be found in power systems. As well as the power components, both analogue and digital electronics form an integral part in turning on and off power semiconductor devices to divert currents at different times to different parts of a power converter circuit. Electronics are also used to measure current, voltage, and power in the control and testing of power electronics. Digital electronics are used to make logical decisions in the operation of a converter and in the implementation of control algorithms. The author hopes that this book reflects his desire to present the fundamental art of power electronics and that the text encourages the reader to seek out and use the latest devices in the design of novel power electronic circuits.

# 1

# Introduction

The subject of power electronics is concerned with solid state devices for the control and conversion of electrical power. These silicon devices are designed mainly for switching, for example, the transfer current from one part of an electrical circuit to another. By switching the devices at a high frequency (e.g., at 100 kHz compared to the normal electrical supply frequencies of 60 or 50 Hz), power electronics can synthesize voltage and current waveforms. These can take the form of well-known functions such as sinusoidal, triangular, ramp, or square, or more complex and variable shapes depending on the load and application.

Power electronics has a wide range of applications from the small systems used in electrical appliances to very large systems for the supply and distribution of electricity. In the lighting industry, a high-voltage pulse is used to start the ionization of the gas in a fluorescent tube and then maintain a steady current at a high frequency once ionization has taken place. An electrical vehicle is propelled forward using power electronics and an electrical machine to convert electrical from its batteries into mechanical energy. On a demand for breaking, the power electronics and electrical machine can be used to slow the vehicle by recovering the kinetic energy stored in the vehicle. Ocean passenger liners typically have several diesel engines or gas turbines as their primary source of power. The mechanical energy from the engines is converted to electrical energy using synchronous machines. Then, through power electronics, the energy is converted back into mechanical energy using another set of electrical machines to propel the ship. The power electronic systems and electrical drives provide an efficient, compact, and flexible solution for maneuvering a ship in a port and sailing in all kinds of weather while traveling the oceans. Power electronics is an essential technology for the conversion of the mechanical power produced by a wind turbine into electrical power and interfacing it to an electrical power system (electrical grid). The variable nature of the rotational speed of a wind turbine demands careful design of the power electronics and control strategies to economically and efficiently generate electricity.

## 1.1  Power Semiconductor Devices

The subject of power electronics relies on knowledge from other disciplines. Mathematics is the cornerstone of any engineering design; for power electronics the solution of differential equations, Laplace transforms and Fourier series are particularly useful. Knowledge of analog electronics, circuit theory, control theory, computer simulation, digital electronics, heat transfer, electromagnetism, and sensors are also important in varying degrees. In any application, these subjects form the basis for the design of a power electronic circuit, which should be viewed as a system where the various components interact both theoretically and practically.

## 1.2  Power Conversion

Electrical supplies can be classified into either alternating current (ac) or direct current (dc). Collectively, power electronic circuits form a power converter that can transfer energy from a supply to a load or an energy storage device. For example, the charging of an automotive 12-V battery (dc) from a 120-V (ac) supply. The input can be either ac or dc as can the output, forming four combinations of energy transfer (Figure 1.1). The supply voltage(s) do not have to be equal and for ac supplies the frequencies on either side of a power converter can be the same or different. It is also possible to combine converters, for example, converting serially from ac to dc and then dc to ac. This design exploits the characteristics of individual converters in which directly converting from ac to ac would be less economical or less efficient.

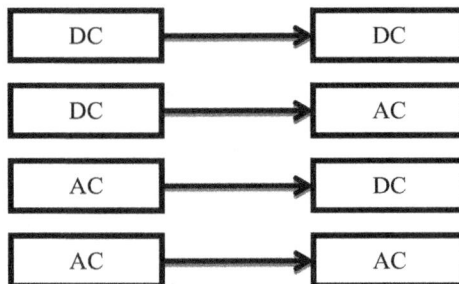

**Figure 1.1**   The four combinations of electrical energy conversion.

A power semiconductor will have parameters that are specified by the manufacturer such as maximum continuous current. By a careful study of device data, a designer can make calculations and select a device or devices that are suitable for the given application. Power can then be transferred smoothly from the input supply to the output load. If electrical energy is stored in load inductors or capacitors, some converters can return energy back to the supply, but others may need extra power components. Besides the power devices, there are other circuits that need the designer's attention (Figure 1.2). Central to the smooth and reliable operation of a converter are control electronics. These circuits may have counters, timers, analog-to-digital converters and memory for look-up tables. Typically, small microcontrollers contain all the necessary functions in one or a few devices. The controller may also be linked to a computer or network. Signals that contain data for the measurement of voltages and currents are sent to the controller to act as inputs to control algorithms running on the main controlling device. This information can also be used to shut down the power circuit should an electrical fault occur or if voltages or currents exceed pre-set thresholds. The electronic circuits need a smooth low voltage supply, which may be 3V, 5V, or 10V depending on the technology. This power can be sourced via a dc-to-dc step-down converter or an ac-to-dc converter depending on the main supply. Parameters such as a set frequency of operation can be sent to the electronic controller and stored or updated as needed. Control algorithms

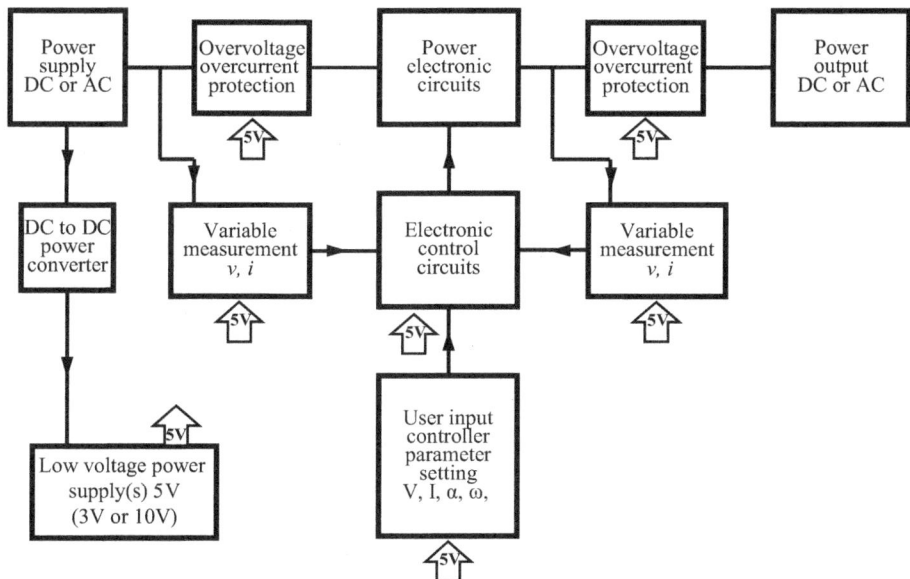

**Figure 1.2** Block diagram of a power converter.

**Table 1.1**
Power Electronic Devices

| | | | |
|---|---|---|---|
| 1. | Bipolar junction transistor (BJT) | 5. | Metal oxide-controlled thyristor (MCT) |
| 2. | Diode | 6. | Metal oxide semiconductor field effect transistor (MOSFET) |
| 3. | Gate turn-off thyristor (GTO) | 7. | Thyristor |
| 4. | Insulated gate bipolar transistor (IGBT) | 8. | Triac |

can be implemented and stable operation of a converter achieved. Over-voltage and over-current components prevent damage to the converter and protect the power semiconductors.

Power converters from a few watts to several megawatts are often coupled with an electric motor. They are called electric drives. The type of power converter depends on the type of rotating machine. Some electric drives require a special power converter for successful operation.

There are different types of devices (Table 1.1). Most of them have three terminals, one of which is a controlling input called a gate or base. A low power signal is applied to this terminal that switches the main output terminals between a conducting state, the on state and a nonconducting state, the off state. Some devices have electronic circuits as well as a power output stage, for example, in the monitoring of current and subsequent shutdown under overload conditions.

## 1.3  Passive Components

Passive components are used in combination for the operation of power semiconductor devices within their specifications. For example, a series-connected resistor and capacitor forming a lowpass circuit is used to limit the rate of change of voltage when a device is nonconducting. The fundamental equations should be familiar to the reader, but are given here as a reminder of their characteristics.

A resistor (resistance, $R$) is governed by Ohm's law.

$$v = R\,i \tag{1.1}$$

Capacitor (capacitance, $C$)

$$v = \frac{1}{C}\int i\,dt \quad i = C\frac{dv}{dt} \quad \Delta I = C\frac{\Delta V}{\Delta t} \tag{1.2}$$

Inductor (inductance, $L$)

$$i = \frac{1}{L} \int v \, dt \quad v = L \frac{dt}{dt} \quad \Delta V = L \frac{\Delta I}{\Delta t} \tag{1.3}$$

For linear rises and falls in current and voltage with respect to time, only the end points are needed in calculations. For example, the capacitance can be determined from the change in current ($\Delta I$) multiplied by the time ($\Delta t$) divided by the voltage ($\Delta V$). In many power electronic circuits, voltages can be assumed to be constant and currents steady, especially when they are varying slowly; for example, the current in an inductor varying in milliseconds can be considered constant when compared to voltage changes occurring in microseconds.

## 1.4  Parameters and Analysis of Waveforms

The maximum permitted currents and voltages of a power semiconductor are stated in data sheets and given in terms of parameter values. When testing a converter design, data are collected and analyzed to see if the devices are not being subjected to values outside their specifications. These data are summarized by calculations of their means, root mean squares and form factors.

The mean (average) of a function, $f(t)$, which has a period, $T$, is

$$f_{mean} = \frac{1}{T} \int_0^T f(t) \, dt \tag{1.4}$$

In general, $f(t)$ does not have to be a periodic function.
The root mean square (RMS) is

$$f_{RMS} = \sqrt{\frac{1}{T} \int_0^T f(t)^2 \, dt} \tag{1.5}$$

The form factor is

$$F_f = \frac{f_{RMS}}{f_{mean}} \tag{1.6}$$

## 1.5  Ideal Power Device

Consider a mechanical switch that is used to turn on and off a room light. When it is conducting current, two metallic contactors are pushed together and held in place by a spring. There should be only a very small resistance between the

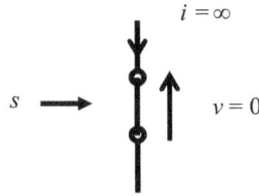

**Figure 1.3**   Ideal switch in the on state.

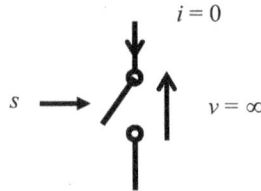

**Figure 1.4**   Ideal switch in the off state.

contacts so that all of the electrical power is delivered to the load (room light). When the contacts are apart, they are separated at a distance so that the resistance created by the air between them is very high and there is no power in the load. There is a limit to the maximum current that can be conducted safely through the metal contacts, which depend on the metal used (copper, silver, or gold), contact area, and switch construction. In the off state, there is a limit to the voltage difference across the contacts before the air breaks down and causes the switch to malfunction. The rate at which the switch can be turned on and off is limited to a low operational frequency.

In comparison, the ideal characteristics of a mechanical switch and a power semiconductor device are shown in Figures 1.3 and 1.4. In its on state, there is no voltage across the device and it can pass an infinite current between its terminals. The transition between on state and off state is achieved instantaneously (zero seconds). Similarly, there is a zero delay in switching between the off and on state. While the device is in its off state, an infinite voltage can exist between its terminals and there is no current flow (zero leakage current).

These ideal characteristics would lead to high speed switching, no power loss, and a simple electronic signal, $s$, to turn on or turn off the device. To change the state of the switch, there is only a transfer of information and no power is consumed in the process.

## 1.6   Practical Device

A power semiconductor device in an electrical circuit will experience variations in voltage and current. It may experience very quick changes (e.g., voltage spikes on mains supplies caused by lightning) or slower changes (e.g., when an electric drive in a factory is accelerated to its operational speed). A device should be selected so that it can withstand all the voltages that are applied to its terminals. Often, the voltage rating is higher than the normal operating voltage so that it can withstand any expected transient higher voltages. Similarly, the current rating should be sufficient for the expected operation so that during faults, surges in current are kept within the rating of the device. A practical device will have voltage across its terminals with current flowing and it will dissipate heat with a higher temperature inside the device compared to the outside environmental temperature. Cooling of the device is usually required to keep the temperature of the silicon to within the limits specified for the device by the manufacturer.

## 1.7  Sources of Information

There are a number of textbooks that cover general concepts about power electronics but there are also books that describe power converters for particular applications. Electric drives are covered extensively in some texts for induction motors, dc motors, synchronous motors, stepper motors, and some less common machines such as the switched reluctance motor. The transmission and distribution of electricity are also important topics as is the interface between a source of power such as a solar cell and an electrical power system. Practical sources of information can be found in the data sheets supplied by the manufacturers of individual devices. Large and small companies, making power converters, also have useful details about power converter concepts. Future technological developments can be found in conference proceedings and journal papers that are published by professional organizations. The Institute of Electrical and Electronic Engineers publish several journals that contain research and development articles in power electronics (e.g., *IEEE Power Electronic Letters* and *IEEE Transactions on Power Electronics*). The Institution of Electrical Technology also publishes journals (e.g., *The IET Power Electronics*). New switching devices require the development of new materials that can be found in physical science conference proceedings and journals.

## Selected Bibliography

Bose, B. K., *Power Electronics and Motor Drives*, Boston, MA: Elsevier Academic Press, 2006.

Davis, R. M., *Power Diode and Thyristor Circuits*, London, UK: Peter Peregrinus, 1976.

Dewan, S. B., G. R. Slemon, and A. Straughen, *Power Semiconductor Drives*, New York: John Wiley & Sons, 1984.

Hart, D. W., *Introduction to Power Electronics*, Upper Saddle River, NJ: Prentice Hall, 1997.

Koss, A., *A Basic Guide to Power Electronics*, New York: John Wiley & Sons, 1984.

Lander, C. W., *Power Electronics*, 3rd ed., New York: McGraw-Hill, 1993.

Mohan, N., T. M. Undeland, and W. P. Robins, *Power Electronics: Converters, Applications, and Design*, 3rd ed., New York: John Wiley & Sons, 2003.

Rashid, M. H., *Power Electronics Handbook: Devices, Circuits, and Applications*, 3rd ed., Boston, MA: Elsevier, 2011.

Sood, V. K., *HVDC and FACTS Controllers: Applications of Static Converters in Power Systems*, Boston, MA: Kluwer Academic Publishers, 2004.

Williams, B. W., *Power Electronics: Devices, Drivers, Applications, and Passive Components*, New York: Macmillan, 1992.

# 2

# Diode

Diodes are manufactured in a wide range of sizes and specifications. The circuit symbol for a diode is shown in Figure 2.1 where current flows from the anode to the cathode when the anode terminal has a positive voltage with respect to the cathode terminal.

A diode allows for current flow in one direction only (forward biased where the anode has a positive voltage with respect to the cathode) and has no current flowing through it when reverse biased. This characteristic of a diode is called rectification. Diodes are manufactured with a specification of an average current in excess of 3,000A when forward biased and 9,000V repetitive peak reverse voltage when reverse biased.

## 2.1 Ideal Characteristic of a Diode

In forward conduction, there is no voltage drop between the anode and cathode and a diode can have infinite current (zero resistance). In reverse blocking, there is no current flow and an infinite voltage can exist between the diode terminals (infinite resistance). This ideal model is useful when thinking about the first ideas in the design of a power electronic circuit as it provides the basic concepts without the details of a practical device (Figure 2.2).

## 2.2 Simple Diode Model

For small positive voltages across the anode and cathode, there is no current flow (Figure 2.3). In forward conduction, there is a constant voltage difference

Anode

Voltage ↑    Current ↓

Cathode

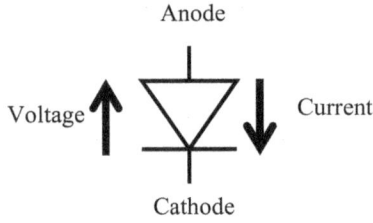

**Figure 2.1**   Diode symbol.

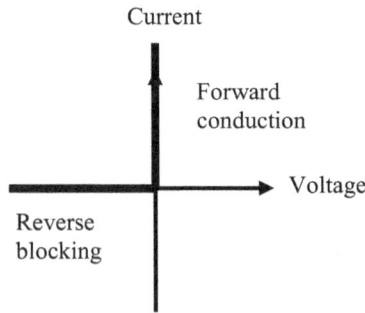

Current

Forward
conduction

Voltage

Reverse
blocking

**Figure 2.2**   Ideal characteristic of a diode.

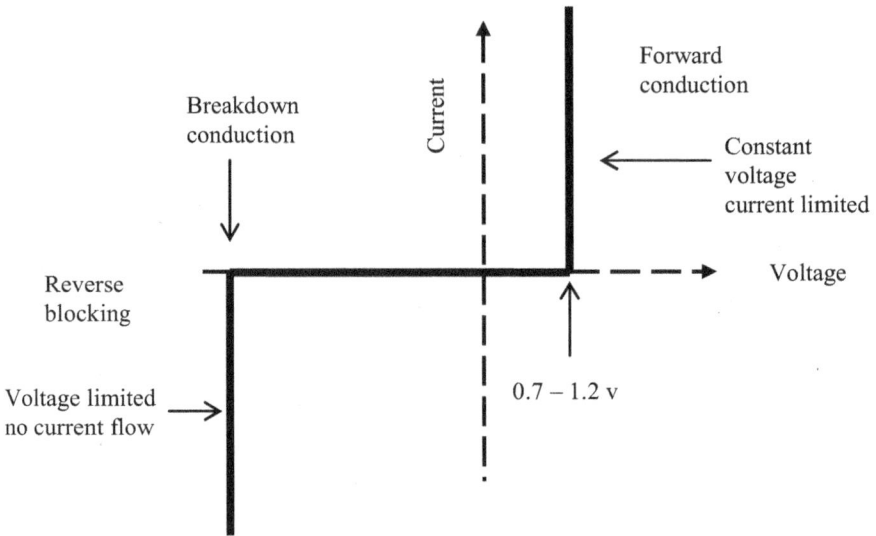

Breakdown
conduction

Current

Forward
conduction

Constant
voltage
current limited

Reverse
blocking

Voltage

Voltage limited
no current flow

0.7 – 1.2 v

**Figure 2.3**   Simple diode model characteristic.

between the anode and cathode (typically about 0.7V for a small electronic diode and about 1.2V for a power diode). It has an infinite current capability. In reverse blocking, there is no current flow until the voltage reaches the breakdown voltage when there is infinite current. The current in the forward direction and the applied voltage in the reverse direction are limited by the characteristics of the individual device. The forward current and the breakdown voltage have a wide range. This model is perhaps the most useful in that it accounts for the overall behavior of a diode and includes the main practical limitations on forward current and reverse voltage.

## 2.3  Diode Model

The voltage and current in a diode are nonlinearly related as shown in

$$i = I_0 \left( e^{\frac{v}{V_T}} - 1 \right) \text{ where } V_T = \frac{nkT}{q} \tag{2.1}$$

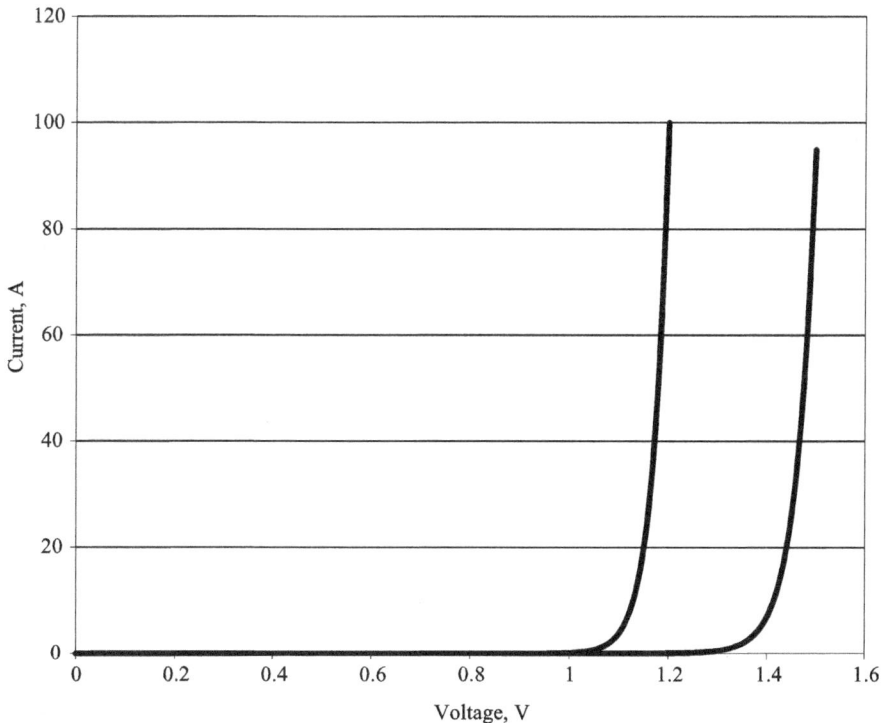

**Figure 2.4**  Diode model for forward biased voltage. On the right side, there is a greater voltage drop as the diode becomes hotter.

where $q$ is the electron charge $(1.6 \times 10^{-19}$ C), $k$ is the Boltzmann constant $(1.38 \times 10^{-23}$ J K$^{-1})$, $T$ is the absolute junction temperature, $I_0$ is the dark saturation current, and $n$ is the parameter that is dependent on the individual device. In forward conduction, the current rises when the voltage reaches about 1V (Figure 2.4). There is a larger voltage across a diode when it becomes hot. In reverse bias, a diode exhibits a leakage current (Figure 2.5) that also increases with increasing current.

Rearranging (2.1),

$$i = I_0 e^{\frac{v}{V_T}} - I_0 \tag{2.2}$$

Neglecting $I_0$ at the end of (2.2) and taking logarithms, then approximately

$$\ln i \approx \ln I_0 + \frac{v}{V_T} \tag{2.3}$$

From which the parameters of the diode can be estimated from values of current and voltage by plotting the logarithm of the current against voltage.

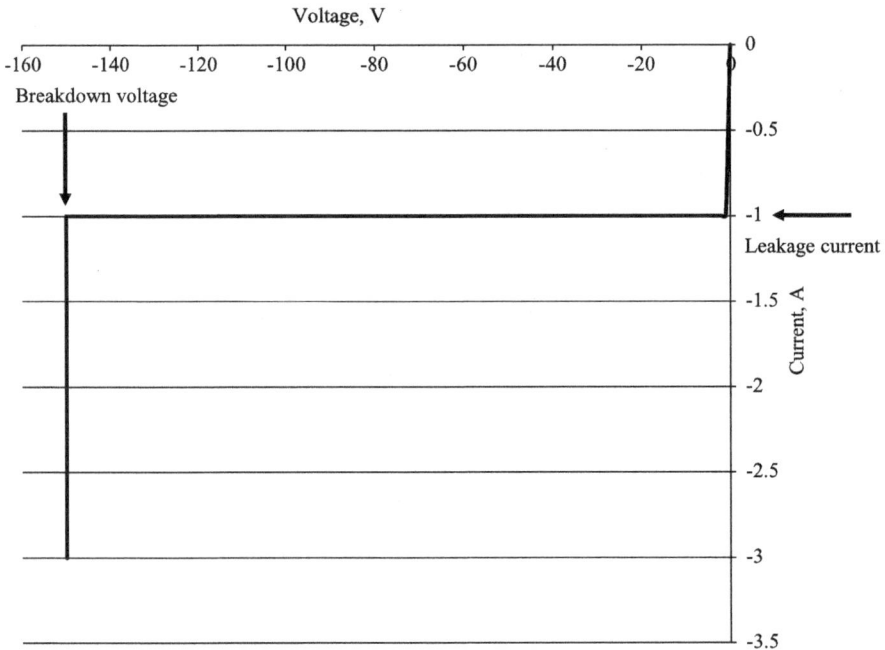

**Figure 2.5**  Diode model for reverse voltage showing leakage current and breakdown voltage.

## 2.4 Power Dissipation

The power dissipated in a diode can be calculated from (2.2) or (2.3). The current flowing through the diode could be constant. In this case, the power is simply calculated by multiplying the resulting voltage [from (2.2) or (2.3) or experimentally measured] by the current. However, if the current is varying, then the calculation requires an approximation for the heat generated, by splitting the current into its average and alternating components (Figure 2.6). During forward conduction, the power dissipated in a diode can be estimated using

$$P = V_f I_{av} + I_{RMS}^2 R_f \tag{2.4}$$

where $V_f$ is the approximate forward voltage, $I_{av}$ is the average current, $I_{RMS}$ is the root mean square current, and $R_f$ is an approximate resistance. The current values are calculated analytically from design waveforms or from measured data. The resistance can be estimated by an inspection of the slope of the diode characteristic. In Figure 2.6, the characteristic is approximately linear from a current of 30A to 100A with a resistance of about 0.48 mΩ (reciprocal of the slope).

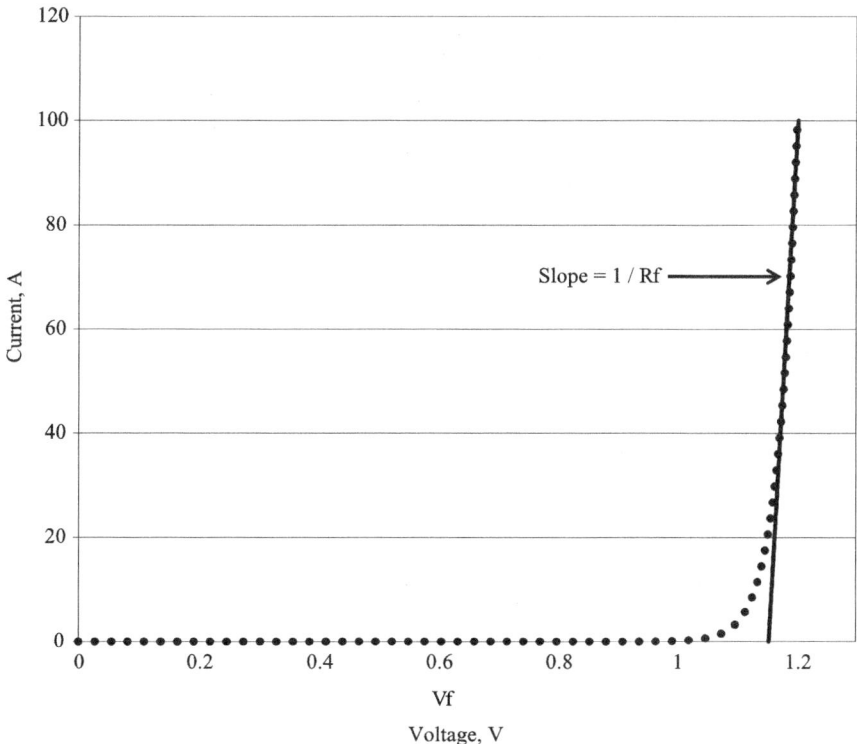

**Figure 2.6** Approximation for the forward conduction characteristic for a diode.

## 2.5  Free-Wheeling Diode (Fly-Wheel)

Consider the circuit in Figure 2.7. If the switch is closed and the initial current is zero in the inductor, then the current, $i$, is

$$v = L\frac{di}{dt} \quad or \quad i = \frac{V}{L}t \tag{2.5}$$

This equation states that the current will continue to increase for all time. In practice, the switch will heat up and eventually be destroyed. If the switch is opened when there is current flowing through the inductor then the voltage, $v_0$, will become very large as the rate of change of current with time is large [see (2.5)]. Again, the switch will be destroyed as the voltage across its terminals will be too large and exceed the rating of the device. The energy stored in the magnetic field of the inductor will be dissipated as heat, light, and sound. A practical application of generating a large and transient voltage, by suddenly interrupting the current in an inductor, is in the spark ignition system of an internal combustion engine where 20 to 30 kV are produced. In other applications, where there is a need to prevent a large voltage appearing across the switch, a diode (freewheeling) is often placed in parallel with the inductor (Figure 2.8).

As the switch is opened, the voltage across its terminals will rise and exceed the supply voltage, $V_d$. At this time, the diode becomes forward biased and will conduct. The voltage, $v_0$, will be $V_d$ plus about 0.7V for a small diode or about 1.1V for a power device.

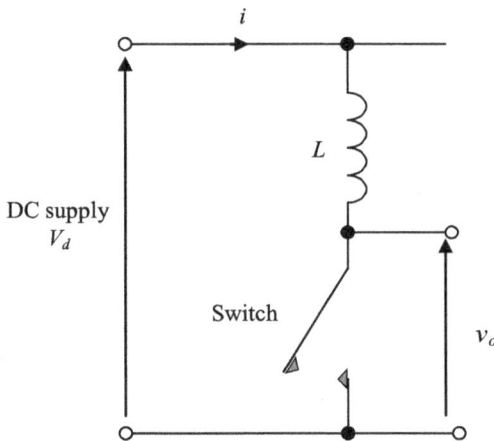

**Figure 2.7**  Inductor switching without a free-wheeling diode.

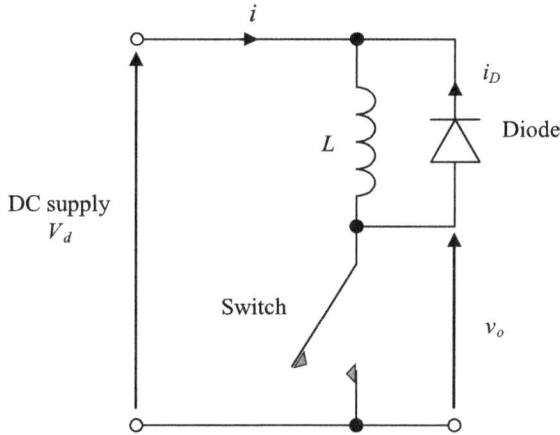

**Figure 2.8**   Inductor switching with a free-wheeling diode.

## 2.6   Zener Diode

The Zener diode is designed to operate at the reverse breakdown voltage in reverse conduction (Figure 2.3). It is used as a voltage reference device and as a voltage clamp. It can also be used with a freewheeling diode. An application of this device is for a solenoid circuit during the switching of a relay. When the switch in the circuit (Figure 2.9) is closed, the current is limited by the resistance of the solenoid coil (not shown in Figure 2.9). On opening the switch, the voltage, $v_o$, across it rises. When voltage across the switch terminals reaches a value

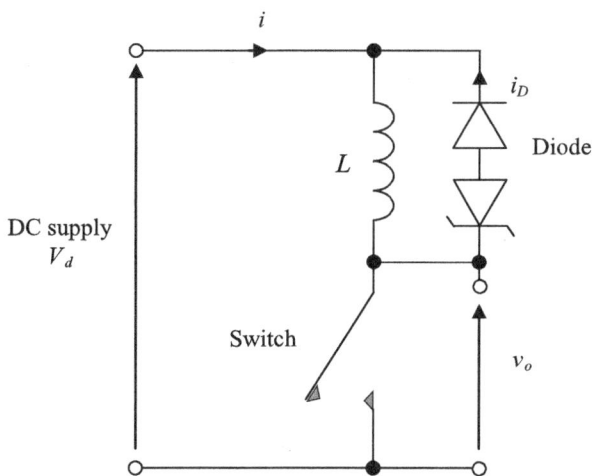

**Figure 2.9**   Switching an inductor with freewheeling diode and Zener diode.

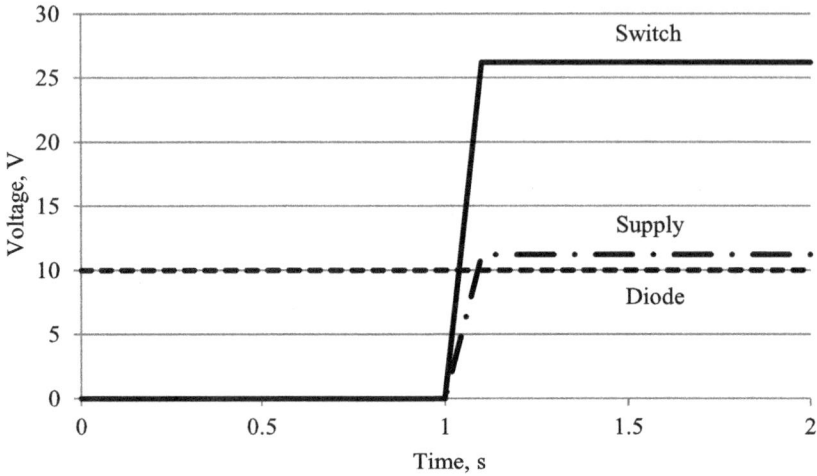

**Figure 2.10**   Waveforms for freewheeling diode and Zener diode circuit.

equal to the combined voltages of the Zener diode, the dc supply, and the diode, the voltage stops rising and remains steady. The voltage across the inductor has been limited to that of the forward diode voltage plus that set by the Zener voltage.

In Figure 2.10 are shown the voltages in the circuit for a supply of 10V. The diode has a forward voltage of 1.2V and the Zener diode has a reference voltage of 15V. The addition of the Zener diode increases the voltage across the inductor when the switch is opened and hence the rate of change of current is higher than with just a diode, leading to a faster fall in current (the power is dissipated in the diodes). Zener diodes can have ratings up to 270V and 75W.

In practice the mechanical switches shown in the figures are transistors which are described and discussed in Chapter 4.

## 2.7   Schottky Diode

These diodes have a very high switching rate (almost no switching loss) and can have a reverse blocking voltage of 1.2 kV, a forward conduction current of 27A and a forward conduction voltage of 1.8V.

## Selected Bibliography

Davis, R. M., *Power Diode and Thyristor Circuits*, London, UK: Peter Peregrinus, 1976.

Lander, C. W., *Power Electronics*, 3rd ed., New York: McGraw-Hill, 1993.

# 3

# Thyristor

The thyristor is a four-layered semiconductor that has three terminals (see Figures 3.1 and 3.2). Current passes from the anode to the cathode when the device is turned on. It is turned on (triggered) by applying a small current, the gate current, through the gate to the cathode (Figure 3.3) and when the anode current is above a specified parameter value called the latching current. On removal of the gate current, a thyristor will stay on if the anode current is above the device's holding current. The current parameters are specified in the data sheet of the manufacturer of the thyristor. Both the latching and holding currents are small when compared to the maximum allowed anode current. The thyristor is not the same as other devices in the sense that the input terminal can be inactive (in the case of the thyristor have no gate current) yet still be on and allow for power to be transferred from one part of a circuit to another. It is only when the anode current falls below the holding current and a gate current is absent that a thyristor will turn off. Typical gate currents are less than 1A.

The gate current is limited by the input impedance between the gate and the cathode terminals. In Figure 3.3, a battery could supply power to the gate and the thyristor triggered by closing the switch. The mechanical switch shown in the circuit could be replaced by a transistor. Notice that the cathode current is the sum of the anode current and gate currents. A typical supply voltage is 10V. Only a simple voltage pulse is needed to turn on a thyristor (Figure 3.4). If the circuit load has a large inductance, then the anode current will rise slowly. Under these circumstances, the gate voltage can be a series of pulses (Figure 3.5). Alternatively, during circuit development, a resistor can be connected in parallel with the inductive load whose value is such that the anode current quickly reaches the latching current.

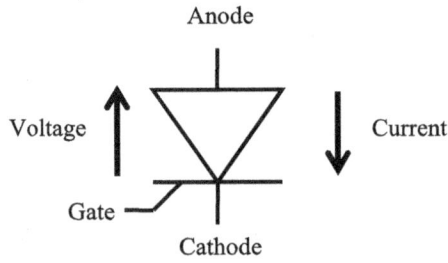

**Figure 3.1**　Thyristor symbol.

**Figure 3.2**　Example of a thyristor packaged as a "hockey puck." The cathode and anode are at the top and bottom. Connections to the gate are made with the two terminals shown on the left of the device.

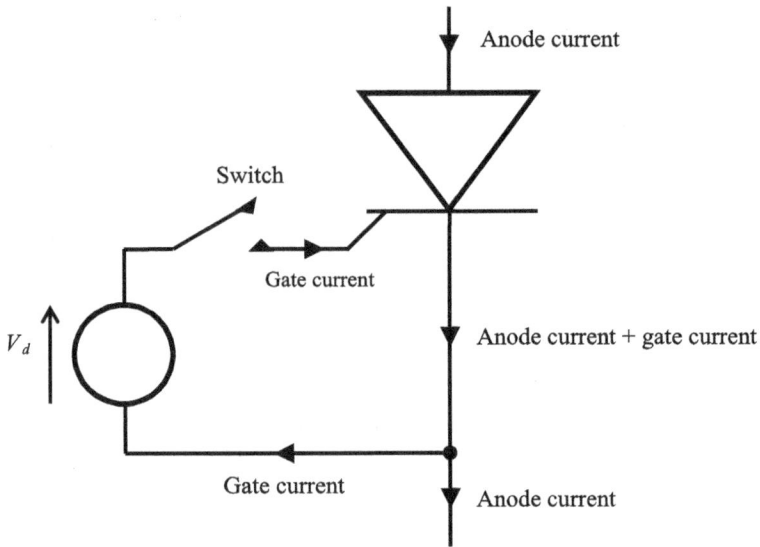

**Figure 3.3**　A simple thyristor gate circuit.

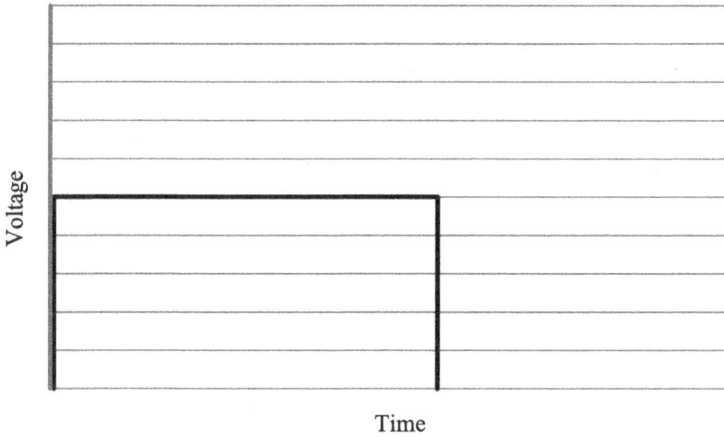

**Figure 3.4**   Typical gate voltage pulse.

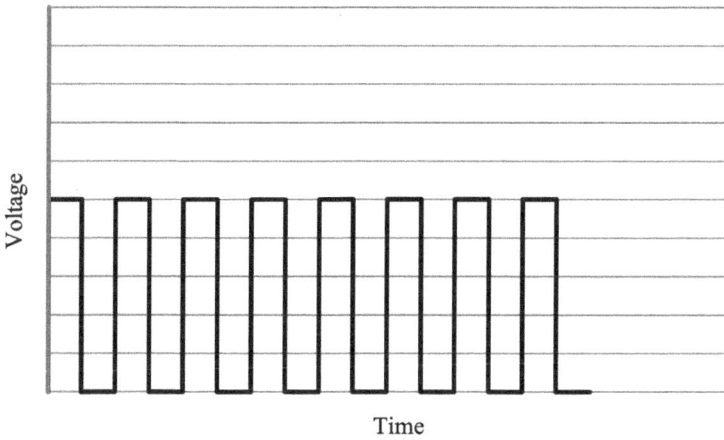

**Figure 3.5**   Gate voltage pulses are used where the load inductance is large and/or the anode current rises slowly ($di/dt = v/L$).

## 3.1   Gate Isolation Circuit

The cathode voltage of a thyristor with respect to a line, neutral or zero (earth) voltage is often several hundreds of volts or more. Information about when to turn on a thyristor is sent from an electronic controller that is in an earthed box. A person can then safely set control parameters without being exposed to dangerous voltages. The signals and power transmitted to a thyristor gate need to be isolated from the earthed control box (Figure 3.6). One way of achieving this aim is to place a 1:1 pulse transformer between the controller and the thyristor gate. The primary voltage, $v_p$, is then referenced with respect to the earth potential

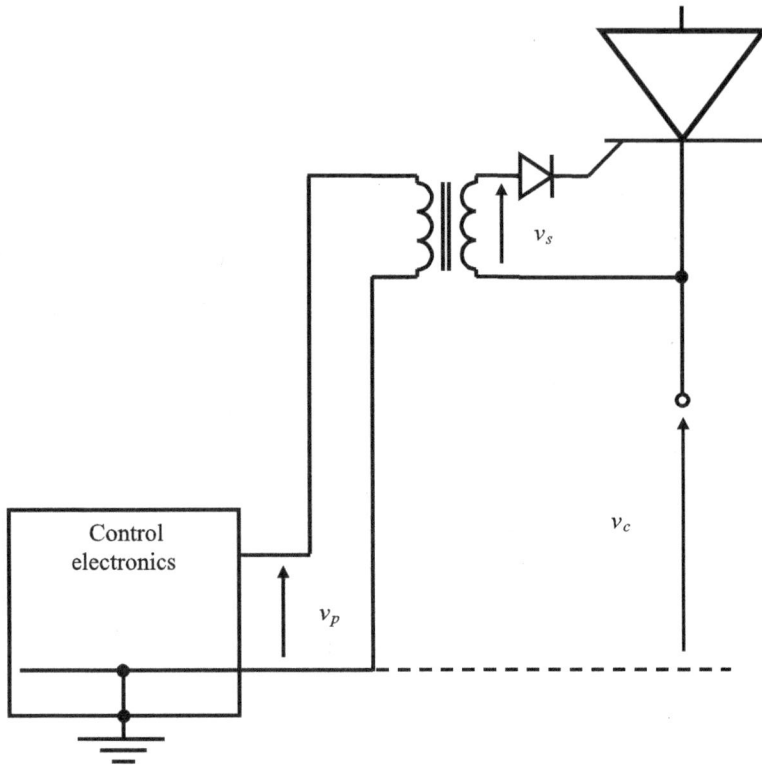

**Figure 3.6**　Pulse isolation circuit.

while the secondary voltage, $v_s$, is referenced with respect to the cathode voltage, $v_c$. During the operation of a converter, the cathode voltage can then vary by several hundreds of volts with respect to the earth potential without harming the control electronics. Another method is to send the signal information using light via a fiber optic cable or opto-isolator. Typically the light is in the infrared spectrum but an isolated dc power source is needed at the gate to provide power for the gate current and electronics.

## 3.2　Snubber

A series resistor and capacitor are connected between the anode and cathode to remove transient voltages while the device is turned off (Figure 3.7). With no gate pulse, if the rate of change in voltage across a thyristor exceeds the maximum specified for a device, it will turn on. This parameter for a thyristor is called the maximum critical rate of rise of off-state voltage ($dv/dt$), which is typically 1,000 $V\mu s^{-1}$. A power supply will have sudden transient voltages that can be removed

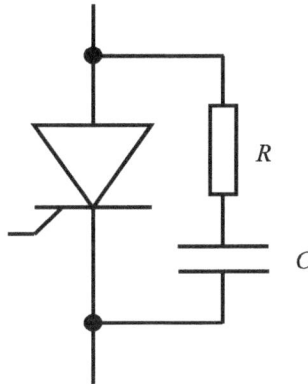

**Figure 3.7** A resistor and capacitor snubber for a thyristor.

at the thyristor terminals by the connection of the capacitor. This effect assumes that there is inductance in series with the thyristor which there typically is from supply cables or transformers. Additional inductance can be added if required. During turn-on, the resistor limits the initial transient current flowing into the thyristor, from the discharge of the capacitor.

## 3.3 Natural Commutation

Thyristors are used mainly in naturally commutated circuits. For a thyristor to turn off, the anode current needs to be taken below the holding current value. Using an ac supply provides the necessary current paths and reverse voltages to reduce the anode current in a natural sequence. Thyristors are only supplied with a trigger pulse at the correct time in sequence and no account need be taken of when to turn off a thyristor.

## 3.4 Single-Phase AC Supply with Single-Gate Pulse Applied to a Thyristor and Resistive Load

From Ohm's law, the current, $i$, and voltage, $v$, are in phase in Figure 3.8. When the current falls below the holding current, the thyristor turns off. This event occurs at 180° ($\pi$ radians).

Firing the thyristor in Figures 3.3 or 3.6 at the start of the positive half-cycle (zero crossing point on the supply when the voltage is rising) results in a positive half-cycle of current (Figure 3.9). Just before the middle of the cycle when the voltage across the thyristor goes below the holding current, the thyristor turns off and there is no current in the negative half-cycle.

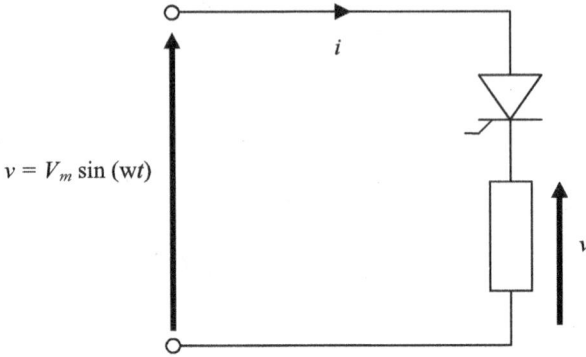

**Figure 3.8**  Single phase supply with a thyristor and resistive load.

Using electronic circuits, the zero crossing point can be detected and a time delay created. Figures 3.9 to 3.12 show the effects on the current waveforms as the delay is progressively increased. The delay is expressed more generally in degrees or radians and is called the delay angle, $\alpha$. The resulting average load voltages and currents are reduced until at an angle of 180° (corresponding to 8.3 ms for a 60-Hz supply or 10 ms for a 50-Hz supply), there is no voltage or current.

The average voltage over a cycle is

$$V_{av} = \frac{1}{2\pi} \int_\alpha^\pi V_m \sin\theta \, d\theta \tag{3.1}$$

$$V_{av} = \frac{V_m}{2\pi} [-\cos\theta]_\alpha^\pi \tag{3.2}$$

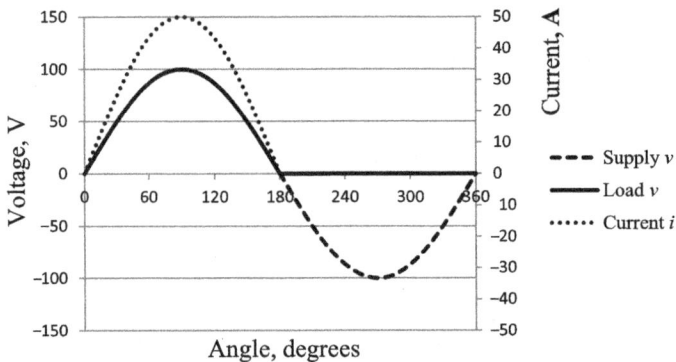

**Figure 3.9**  A 0° delay angle, $R = 2\Omega$.

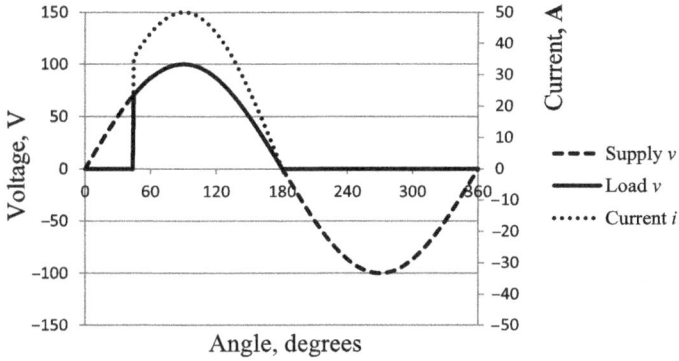

**Figure 3.10**  A 45° (π/4) delay angle $R = 2\Omega$.

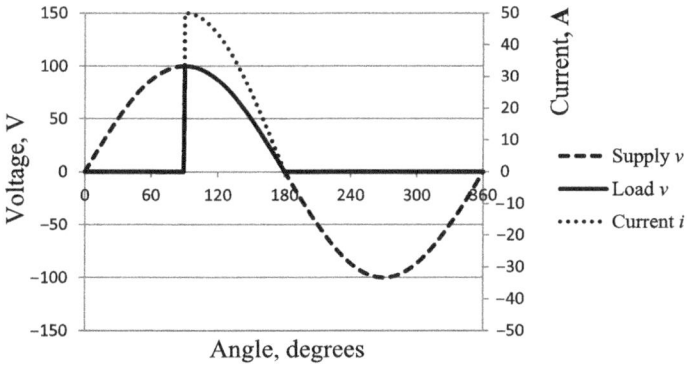

**Figure 3.11**  A 90° (π/2) delay angle $R = 2\Omega$.

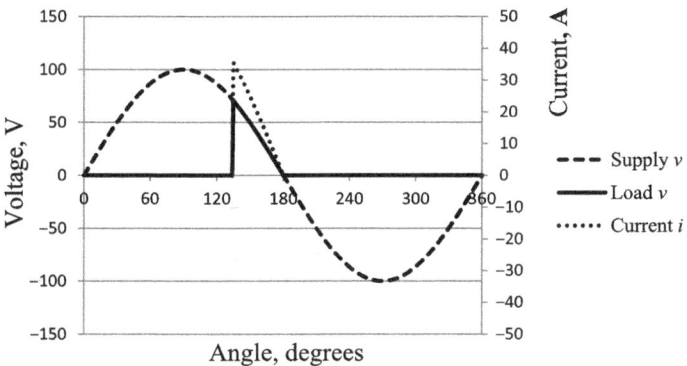

**Figure 3.12**  A 135° (3π/4) delay angle $R = 2\Omega$.

$$V_{av} = \frac{V_m}{2\pi}[1 + \cos\alpha] \qquad (3.3)$$

As the load is a resistor, the average current has the same functional form as the voltage.

## 3.5 Single-Phase Supply with Single-Gate Pulse Applied to a Thyristor and Inductance Load

Applying a single trigger pulse to the gate of the thyristor shown in Figure 3.13 produces one cycle of current that is 90° out of phase with the voltage (Figure 3.14). At an angle just before 360°, the current falls below the holding current and the thyristor turns off. The time (angle) at which the current appears in the inductance can be delayed (Figures 3.15 through 3.20). In these figures there are parts of sine waves of current. In all of these cases, the voltage waveforms are symmetrical about the horizontal axis and at an angle of 180° about the vertical axis. The current waveforms are all positive. Over the positive half-cycle of the supply voltage, energy flows from the supply into the inductor and then during the negative half-cycle energy flows from the inductor to the supply. The direction of energy flow is shown during the positive half-cycle as positive voltage and current while during the negative half-cycle as negative voltage and positive current. Note that there is no loss of energy since it is stored in the inductor and then returned to the supply. At the peak of current when the angle on the horizontal axis is 180°, the rate of change of current is zero corresponding to zero voltage as dictated by the inductor equation.

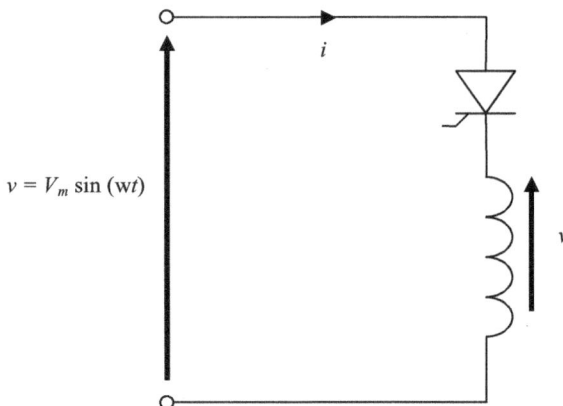

**Figure 3.13** Thyristor and inductor supplied by a single-phase supply.

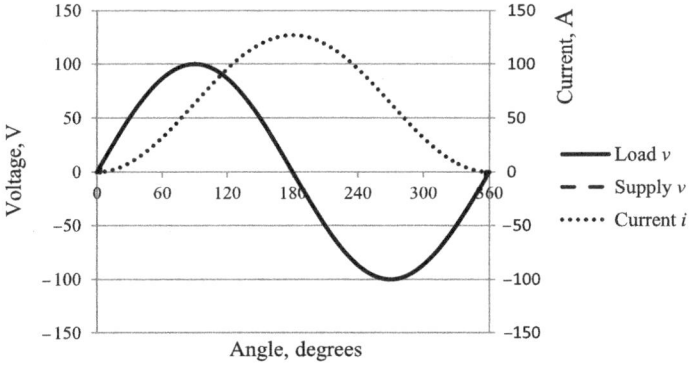

**Figure 3.14**  Single phase with a delay angle of 0° with zero initial current.

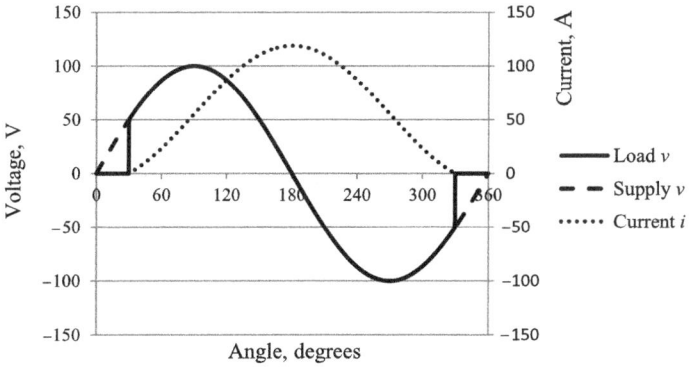

**Figure 3.15**  Single phase with a delay angle of 30°.

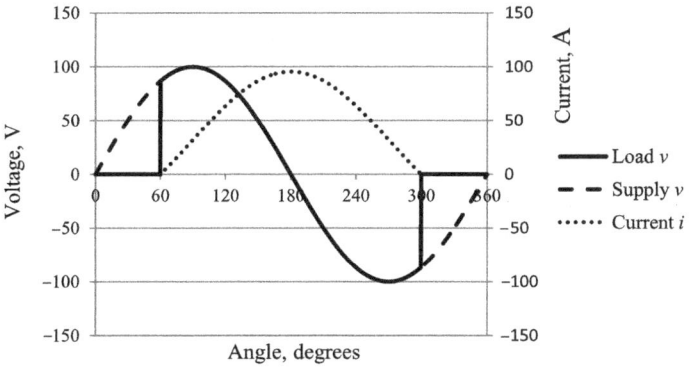

**Figure 3.16**  Single phase with a delay angle of 60°.

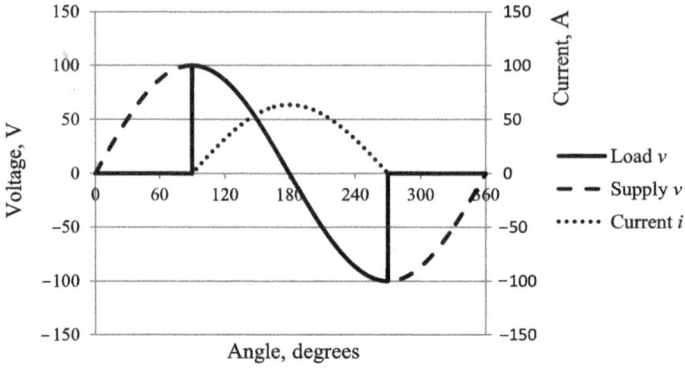

**Figure 3.17**   Single phase with a delay angle of 90°.

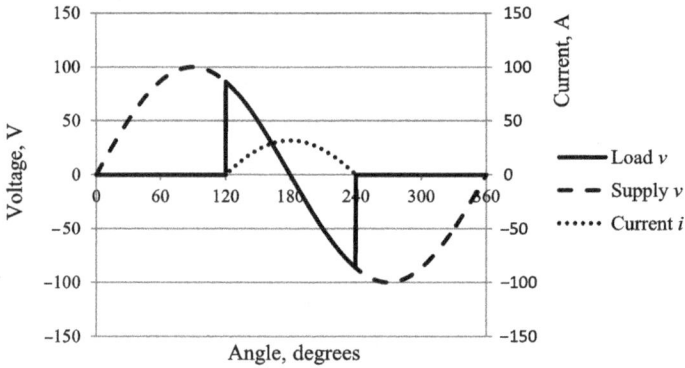

**Figure 3.18**   Single phase with a delay angle 120°.

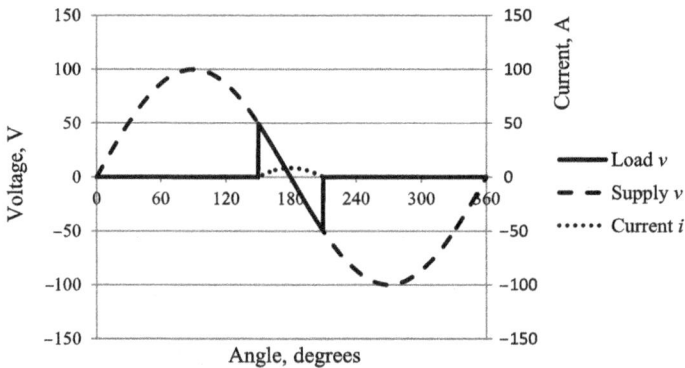

**Figure 3.19**   Single phase with a delay angle of 150°.

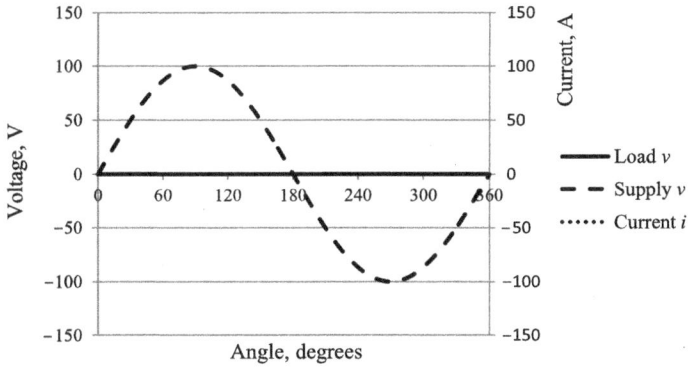

**Figure 3.20**   Single phase with a delay angle of 180°.

For a load which is an inductor, the current is given by

$$V_m \sin \omega t = L \frac{di}{dt} \tag{3.4}$$

$$\int di = \int \frac{V_m}{L} \sin \omega t \; dt \tag{3.5}$$

Integrating this equation

$$i = -\frac{V_m}{\omega L} \cos \omega t + k_3 \tag{3.6}$$

where $k_3$ is a constant.

The boundary conditions are that the current rises from zero at $\omega t = \alpha$. Hence,

$$k_3 = \frac{V_m}{\omega L} \cos \alpha \tag{3.7}$$

$$i = \frac{V_m}{\omega L} (\cos \alpha - \cos \omega t) \tag{3.8}$$

Because the thyristor ceases to conduct current below the holding current, this can be taken as zero and then the equation for current ranges over the angles given by

$$\alpha \le \omega t \le (2\pi - \alpha) \tag{3.9}$$

The current given by (3.8) and (3.9), for various delay angles, is shown in Figures 3.14 to 3.20.

The average voltage over a cycle is zero as the voltage waveforms in Figures 3.14 to 3.20 are symmetrical about the angle of 180°.

The current waveforms, shown in Figures 3.14 to 3.20, are given by (3.8).

Replacing $\omega t$ by $\theta$ in this equation, the average current is given by

$$I_{av} = \frac{V_m}{2\pi\omega L} \int_{\alpha}^{(2\pi-\alpha)} (\cos\alpha - \cos\theta)\, d\theta \tag{3.10}$$

$$I_{av} = \frac{V_m}{2\pi\omega L} [\theta\cos\alpha - \sin\theta]_{\alpha}^{(2\pi-\alpha)} \tag{3.11}$$

$$I_{av} = \frac{V_m}{2\pi\omega L} [(2\pi - \alpha)\cos\alpha - \sin(2\pi - \alpha) - (\alpha\cos\alpha - \sin\alpha)] \tag{3.12}$$

$$I_{av} = \frac{V_m}{2\pi\omega L} [2\pi\cos\alpha - 2\alpha\cos\alpha + 2\sin\alpha] \tag{3.13}$$

$$I_{av} = \frac{V_m}{2\pi\omega L} [2\pi\cos\alpha - 2\alpha\cos\alpha + 2\sin\alpha] \tag{3.14}$$

$$I_{av} = \frac{V_m}{\pi\omega L} [(\pi - \alpha)\cos\alpha + \sin\alpha] \tag{3.15}$$

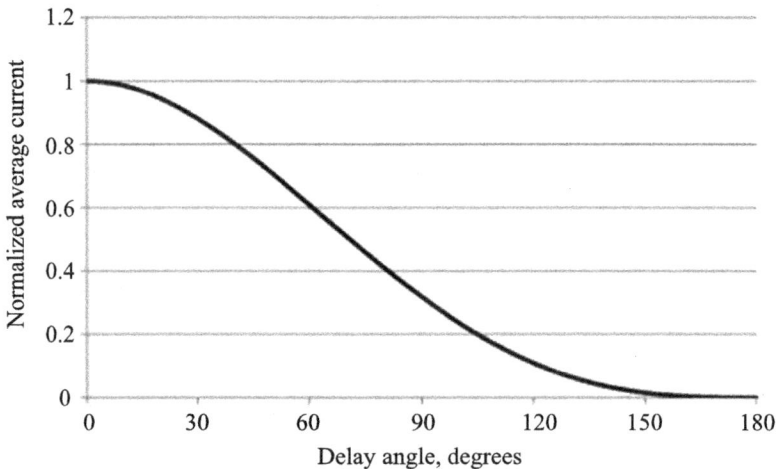

Figure 3.21   Normalized average current as a function of the delay angle, $\alpha$.

The average current (normalized to unity when $\alpha = 0$) is shown plotted in Figure 3.21.

## 3.6 Forced Commutation

If the supply is dc, then the anode current has to be diverted away from the thyristor in order to turn it off. This method of commutation is called forced commutation. It is achieved using a charged capacitor (note the polarity in Figure 3.22) that is placed in parallel with the anode and cathode.

If the switch is closed, the capacitor is connected in parallel with the thyristor and the anode current falls as the capacitor is discharged. The capacitor will charge in the opposite polarity to the supply voltage, $V_d$. If the initial voltage is $-V_d$ on the capacitor, then the change in capacitor voltage is $2V_d$. A load resistor, $R$, will result in an exponential rise in voltage with a time constant of RC.

The switch is typically another thyristor called an auxiliary thyristor (Figure 3.23). To charge the capacitor, a resonant reversal circuit is used. This circuit consists of an inductor in parallel with the capacitor. Without the diode, the current will oscillate forever. The presence of the diode causes the voltage on the capacitor to swing from $V_{cap}$ to $-V_{cap}$. The capacitor current waveform is sinusoidal for half a cycle.

The equation governing the resonance is second order.

$$L\frac{di}{dt} + \frac{1}{C}\int_0^\infty i\,dt = 0 \tag{3.16}$$

The frequency of oscillation is

$$\omega = 2\pi f = \frac{1}{\sqrt{LC}} \tag{3.17}$$

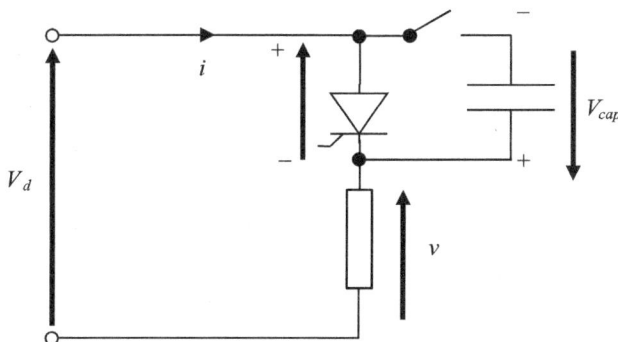

**Figure 3.22** Charged capacitor ready to turn off a conducting thyristor.

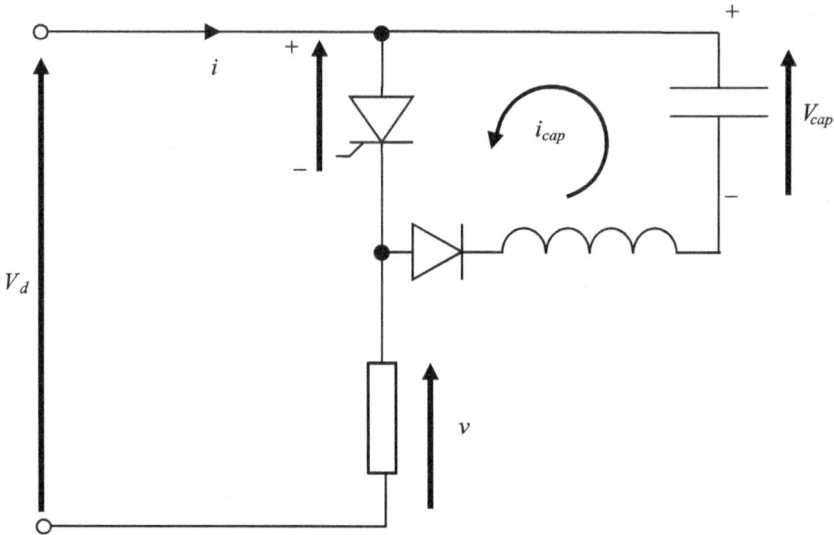

**Figure 3.23** Turn-off of a thyristor using a capacitor, inductance, and diode.

The parallel combination of an inductor and capacitor, forming a resonant circuit, is a useful building block for power converters. Today the thyristor is not used in forced commutation applications as there are better devices for high-power systems where repeated switching is needed. However, this technology is used effectively where there is a single pulse of current into a load.

## 3.7  Triac

This device consists of two thyristors that are connected back to back (in parallel) with one gate input. They are used in static relays, heating regulation, induction motor starting circuits or for phase control operation in light dimmer and motor speed controllers. The load voltage is controlled by varying the delay angle (Figures 3.9 through 3.12). The second thyristor is used to control the voltage during the negative half-cycle that forms part of the triac package (Figure 3.24). In light dimming circuits, the delay is created by the time constant of a resistor and capacitor in series. In other circuits the delay is produced by a digital system using counters.

## 3.8  Gate Turn-Off Thyristors

The turn-on of a gate turn-off thyristor (GTO) requires a pulse of current applied to the gate and is the same method as used with a thyristor. However, it is useful to include an extra high pulse of current into the gate at the start of

Anode 1

Gate

Anode 2

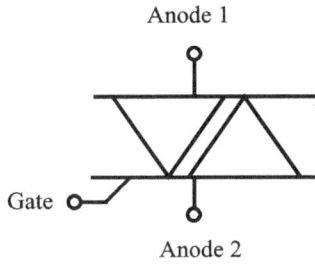

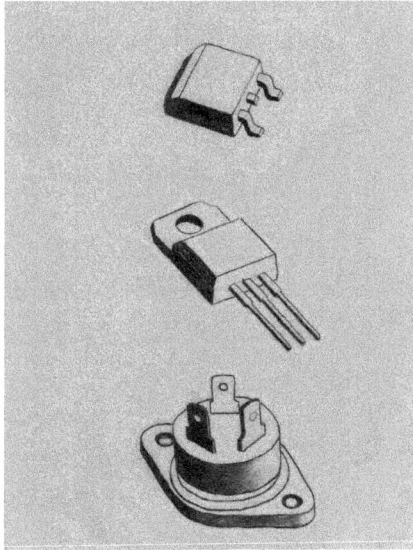

Figure 3.24   The Triac symbol is shown at the top followed by types of device packages that are $D^2$-PAK, TO-220, and RD91.

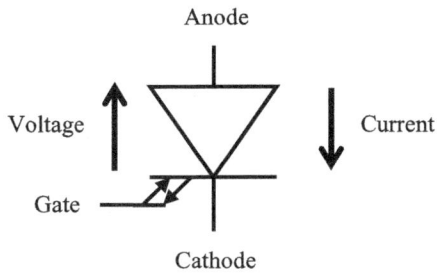

Anode

Voltage

Current

Gate

Cathode

Figure 3.25   Symbol for a gate turn-off thyristor.

turn-on. Turn-off requires the application at the gate of a reverse voltage source with a high current capability (note the arrows in gate of the circuit symbol shown in Figure 3.25). As the name of this thyristor implies, turn-off is controlled at will by the application of electrical power at the gate that does not rely on reducing the anode current to below the holding current threshold before turn-off occurs.

# Selected Bibliography

Davis, R. M., *Power Diode and Thyristor Circuits*, London: Peter Peregrinus, 1976.

Lander, C. W., *Power Electronics*, 3rd ed., New York: McGraw-Hill, 1993.

# 4

# Transistors

The transistor is a three-terminal device (Figure 4.1). There are npn and pnp types. A transistor will be conducting current between its collector and emitter when there is a current flowing into the base and out of the emitter. The voltage across the base to emitter is typically that of a forward-biased diode.

The device currents are related by the following equations (Figure 4.2).

$$i_E = i_C + i_B \tag{4.1}$$

$$i_C = \beta \, i_B \tag{4.2}$$

A small signal transistor has a typical $\beta$ of 100, whereas a power transistor has a value of 20. If $I_c$ is 10A, then with a $\beta$ of 20 there will be a base current of 0.5A.

## 4.1   Safe Operating Area

When a transistor is off, typically there will be the supply voltage across the collector and emitter terminals (lower right in the plot of Figure 4.3). When it is turned on, there will be the saturation voltage and current flowing through its terminals (upper left in the plot). In some circumstances when switching, a transistor will experience both voltage and current and more power dissipation (top right of the plot). However, it can only survive in this state for limited times. This physical phenomenon leads to a safe operating area plot for pulsed operation.

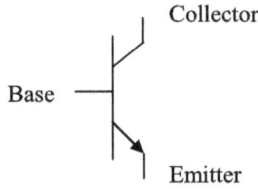

**Figure 4.1**  Symbol for an npn transistor.

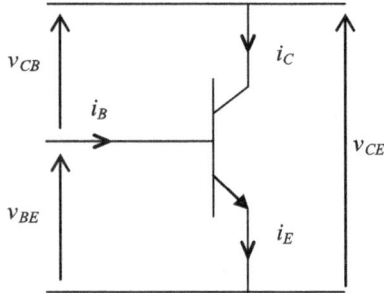

**Figure 4.2**  Voltages and currents in an npn transistor.

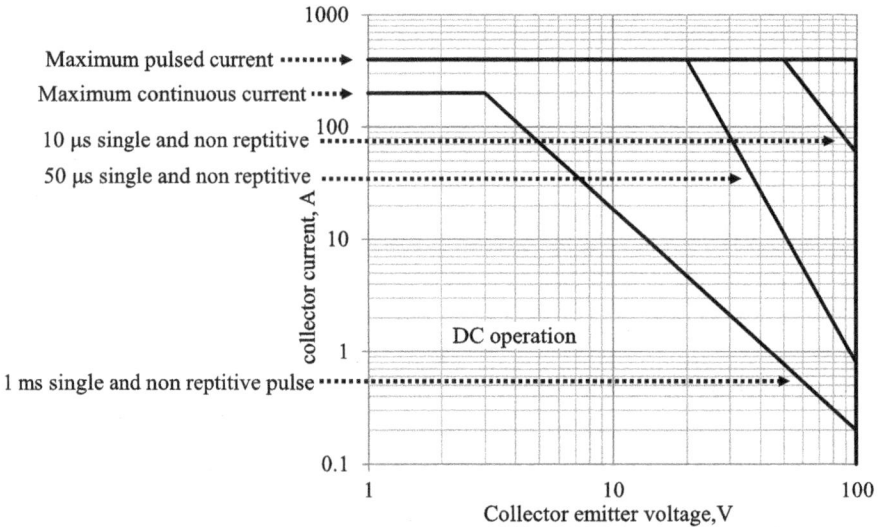

**Figure 4.3**  Safe operating area for an npn transistor.

The boundaries of the plot are determined by peak voltage, peak current, and power rating. Secondary breakdown occurs if the device is taken over the boundaries caused by excessive local heating and high current density. High temperatures increase the electrical conductivity of the semiconductor, resulting in more current, which leads to an increase in temperature and thermal runaway.

## 4.2 Snubber

The turn-off switching characteristic of transistors is improved with the use of a resistor-capacitor-diode snubber. In a typical power electronic circuit, the load is mainly inductive. Consider the circuit with a free-wheeling diode (Figure 4.4). During switching, the inductor current is varying slowly (may be milliseconds) compared to the switching (microseconds) and can assumed to be constant (Figure 4.5). In the turn-off process, the current in the inductor is transferred to the diode and falls linearly with time (top plot in Figure 4.6). The voltage across the capacitor and the collector-emitter rises nonlinearly with time.

Three cases can be considered that depend on the size of the capacitance (Figure 4.7). In the first case, for a small capacitance, the collector-emitter voltage reaches the supply voltage, $V_d$, before all of the inductor current has reached zero. The second case is where the inductor current is zero at the same time that the collector-emitter volatage reaches the supply voltage. In the third case, the

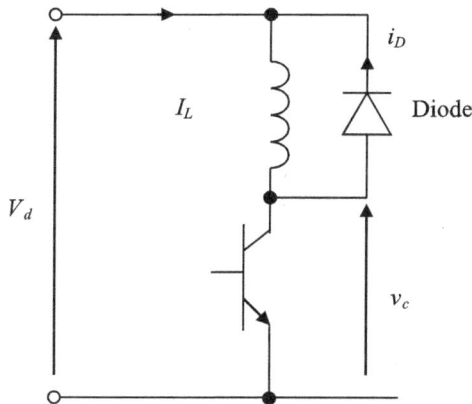

**Figure 4.4** Circuit with a freewheeling diode.

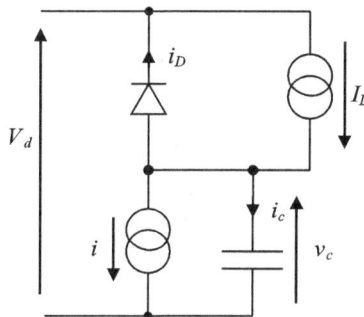

**Figure 4.5** Circuit analysis with an RDC snubber.

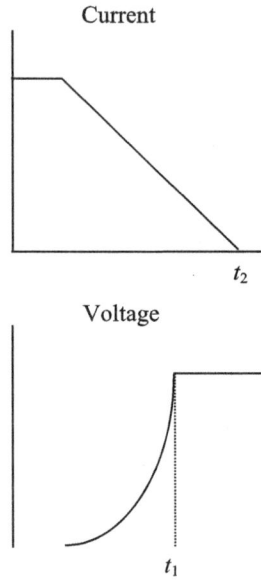

**Figure 4.6**   Linear fall in current and power rise in voltage showing the times $t_1$ and $t_2$.

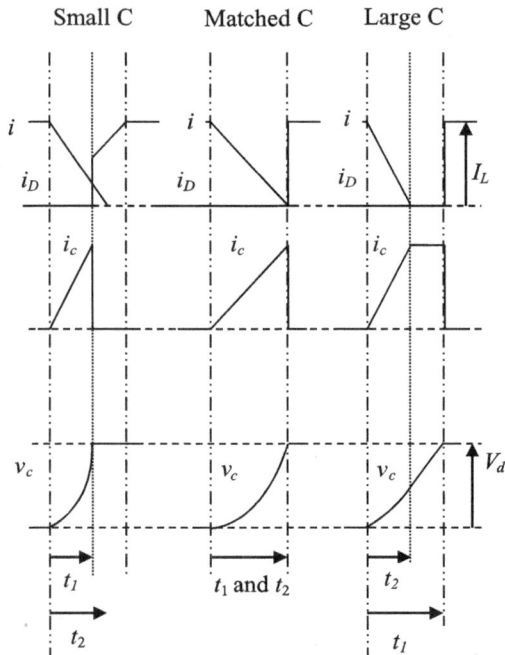

**Figure 4.7**   Current and voltage waveforms for the three cases of small, matched, and large capacitor.

capacitance is large and the collector-emitter voltage reaches the supply voltage at a time after the inductor current is zero.

The energy loss during turn-off can be expressed in terms of the electrical parameters of voltage and current and times taken for the process.

$$energy = \int power \; dt \qquad (4.3)$$

The energy dissipated during the time that the current falls is given by

$$energy = \int_0^{t_2} v_{CE} \; i_C \; dt \qquad (4.4)$$

This equation can be split into the energy while the collector voltage is rising (from zero to time $t_1$) and when the collector voltage is constant and equal to the supply voltage (from $t_1$ to $t_2$).

$$energy = \int_0^{t_1} v_{CE} \; i_C \; dt + \int_{t_1}^{t_2} v_{CE} \; i_C \; dt \qquad (4.5)$$

$$E = E_1 + E_2 \qquad (4.6)$$

Determine the second energy, $E_2$, first. Assuming that the collector current ramps down linearly with time from the load current, $I_L$, to zero in a time, $t_2$.

$$i_C - I_L \left( 1 - t/t_2 \right) \qquad (4.7)$$

Consider the second integral in (4.5) where $v_{CE}$ is the constant supply voltage, $V_d$.

$$E_2 = \int_{t_1}^{t_2} V_d \; I_L \left( 1 - t/t_2 \right) dt \qquad (4.8)$$

$$E_2 = V_d \; I_L \left( t - t^2/2t_2 \right) \Big|_{t_1}^{t_2} \qquad (4.9)$$

$$E_2 = V_d \; I_L \left( t_2 - t_2^2/2t_2 \right) - V_d \; I_L \left( t_1 - t_1^2/2t_2 \right) \qquad (4.10)$$

$$E_2 = V_d \; I_L \left( t_2 - \frac{t_2^2}{2t_2} - t_1 + \frac{t_1^2}{2t_2} \right) \qquad (4.11)$$

$$E_2 = V_d \; I_L \left( \frac{t_2}{2} - t_1 + \frac{t_1}{2t_2} \right) \qquad (4.12)$$

Next, determine the first integral for $E_1$ in (4.5).

The capacitor current is rising from zero as the collector current falls.

$$i_{cap} = \frac{I_L t}{t_2} \tag{4.13}$$

The capacitor voltage is given by

$$v_{CE} = v_{cap} = \frac{1}{C} \int i_{cap} \, dt \tag{4.14}$$

$$v_{cap} = \frac{1}{C} \int \frac{I_L t}{t_2} \, dt \tag{4.15}$$

$$v_{cap} = \frac{I_L t^2}{2 C t_2} \tag{4.16}$$

Substituting (4.7) and (4.16) (the capacitor voltage is equal to the collector-emitter voltage) into the expression for $E_1$ shown in (4.5),

$$E_1 = \int_0^{t_1} \frac{I_L t^2}{2 C t_2} I_L \left(1 - \frac{t}{t_2}\right) dt \tag{4.17}$$

$$E_1 = \frac{I_L^2}{2 C t_2} \int_0^{t_1} t^2 - \frac{t^3}{t_2} \, dt \tag{4.18}$$

$$E_1 = \frac{I_L^2}{2 C t_2} \left[\frac{t^3}{3} - \frac{t^4}{4 t_2}\right]_0^{t_1} \tag{4.19}$$

$$E_1 = \frac{I_L^2}{2 C t_2} \left[\frac{t_1^3}{3} - \frac{t_1^4}{4 t_2}\right] \tag{4.20}$$

During the time that is less than $t_1$, the current is rising and is given by (4.13). At $t = t_1$ the collector-emitter voltage has reached the supply voltage, $V_d$

$$v_{CE} = v_{cap} = Vd = \frac{1}{C} \int_0^{t_1} \frac{I_L t}{t_2} dt \tag{4.21}$$

$$V_d = \frac{I_L t_1^2}{2 C t_2} \tag{4.22}$$

$$\frac{V_d}{t_1^2} = \frac{I_L}{2 C t_2} \tag{4.23}$$

Substituting (4.23) into (4.20) for $E_1$,

$$E_1 = \frac{V_d}{t_1^2}\left[\frac{t_1^3}{3} - \frac{t_1^4}{4t_2}\right] \tag{4.24}$$

$$E_1 = V_d I_L\left[\frac{t_1}{3} - \frac{t_1^2}{4t_2}\right] \tag{4.25}$$

Combining (4.12) and (4.25), the total energy, $E$, is given by

$$E = V_d I_L\left[\frac{t_1}{3} - \frac{t_1^2}{4t_2}\right] + V_d\ I_L\left(\frac{t_2}{2} - t_1 + \frac{t_1}{2t_2}\right) \tag{4.26}$$

$$E = V_d I_L\left[\frac{t_2}{2} - \frac{2t_1}{3} + \frac{t_1^2}{2t_2}\right] \tag{4.27}$$

This equation expresses the switching loss that is dependent on the supply voltage and inductor current as well as the times for the voltage to rise to the supply voltage and for the collector current to fall to zero (Figure 4.8).

Consider the case when time at which the collector-emitter voltage reaches the supply voltage, $V_d$, is equal to the time at which the current is zero, when $t_1 = t_2$.

$$E = \frac{V_d I_L t_1}{3} \tag{4.28}$$

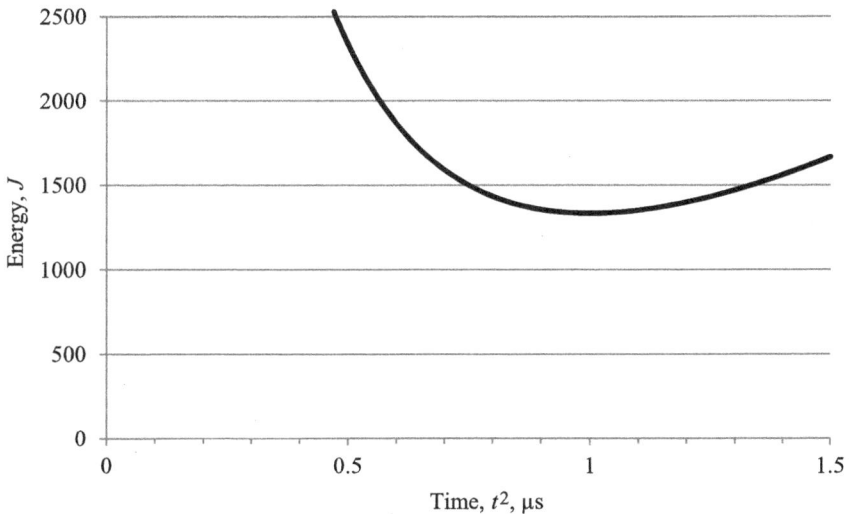

**Figure 4.8**  Energy loss as a function of time, $t_2$ for a supply voltage of 400V and current of 10A with a time, $t_1$ of 1 μs.

The energy is a minimum when the times are equal (Figure 4.8). This can be shown analytically by differentiating partially (4.27) with respect to $t_2$.

## 4.3  Metal Oxide Semiconductor Field Effect Transistor

There are n-channel and p-channel types of metal oxide semiconductor field effect transistors (MOSFET). N-channel devices range from 1.5 kV and 8A to 24V and 429A. P-channel devices range from 250V and 4A to 80V and 110A. MOSFETs are three terminal devices as shown in the circuit symbols (Figures 4.9 and 4.10). Notice that there is a diode (body diode) that is parallel to the drain and source. In a circuit diagram the diode is often not included.

The input to a MOSFET is the application of a voltage between the gate and the source. The output is drain current and drain-source voltage.

The typical output characteristics are shown in the plots of drain current against drain-source voltage (Figures 4.13, 4.14, and 4.15).

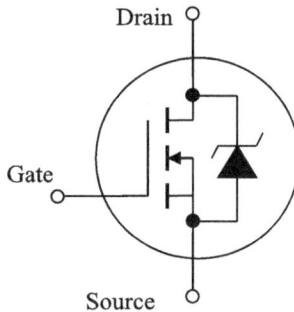

**Figure 4.9**   Circuit symbol for an n-channel MOSFET.

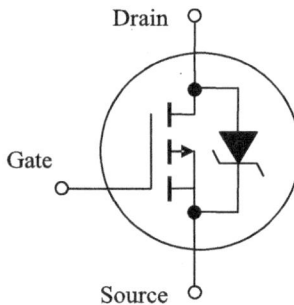

**Figure 4.10**   Circuit symbol for an p-channel MOSFET.

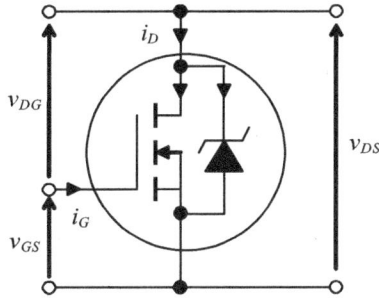

**Figure 4.11**  Voltages and currents of an n-channel MOSFET.

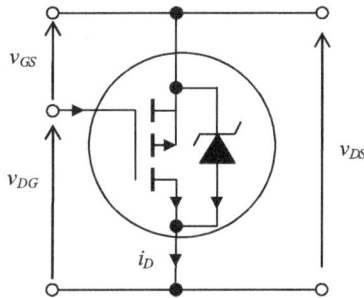

**Figure 4.12**  Voltages and currents of an p-channel MOSFET.

There is a triode region at low drain-source voltages and a saturation region for the higher ranges of currents. The plots from a practical device show small increases in current as the drain-source voltage increases for a constant gate-source voltage in the saturation region. If a MOSFET is off, then the gate-source voltage is zero and the output characteristic follows the horizontal axis. Increasing the gate-source voltage causes drain current to flow and the device is turned on. In power converters a transistor is either off and no current is flowing (bottom right of the output characteristic in Figure 4.13) or there is load current and the transistor is on (top left of Figure 4.13).

The current in the drain terminal is the sum of the MOSFET current and body diode. When the drain-source voltage is positive, the diode is reverse biased and only contributes a leakage current that can be ignored. In most circumstances, when the drain-source voltage is negative, the diode is forward biased while the MOSFET is not conducting and is off. It is also possible with a negative drain-source voltage to turn the MOSFET on by applying a positive gate-source voltage. Under these circumstances, both the diode and MOSFET have current flow but the MOSFET is dominant. This operation lowers the drain-source voltage and hence the power loss. However, to implement this strategy requires coordination of the switching sequence so that the appropriate

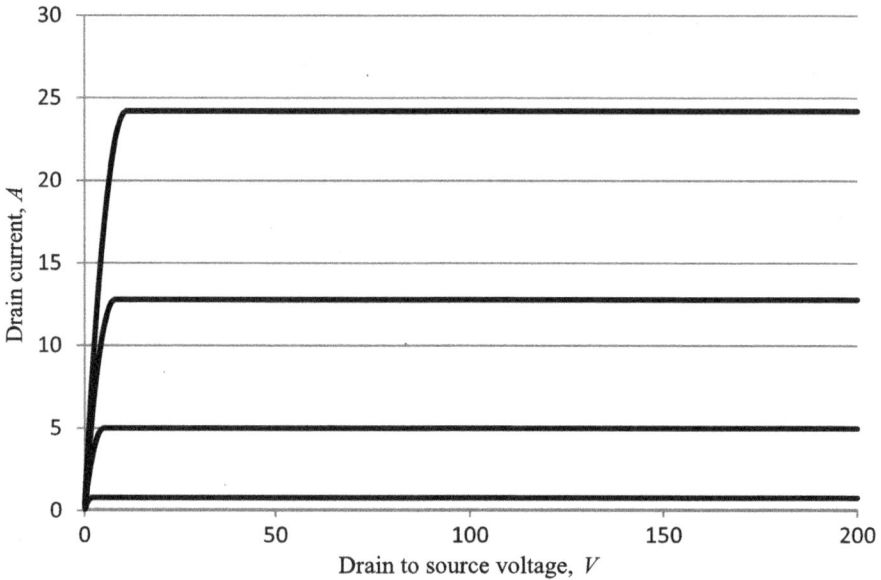

**Figure 4.13** Drain current against drain-source voltage plotted on linear scales for gate-source voltages of 6, 9, 12, and 15V (the curved triode regions are shown to the left of the plot below 10V and the saturation regions are shown horizontally and are independent of the current).

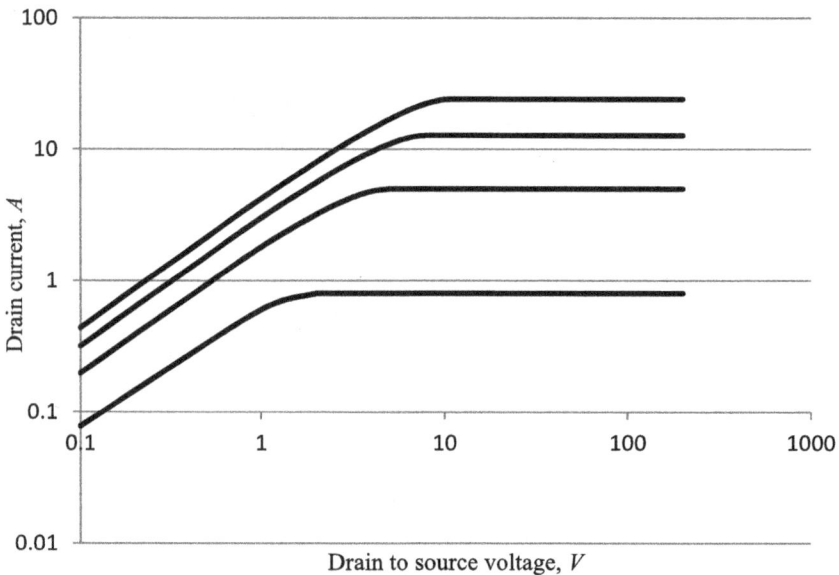

**Figure 4.14** Drain current against drain-source voltage plotted on logarithmic scales, for gate-source voltages of 6, 9, 12, and 15V (the triode regions are shown to the left of the plot below 10V and the saturation regions are shown horizontally and are independent of current).

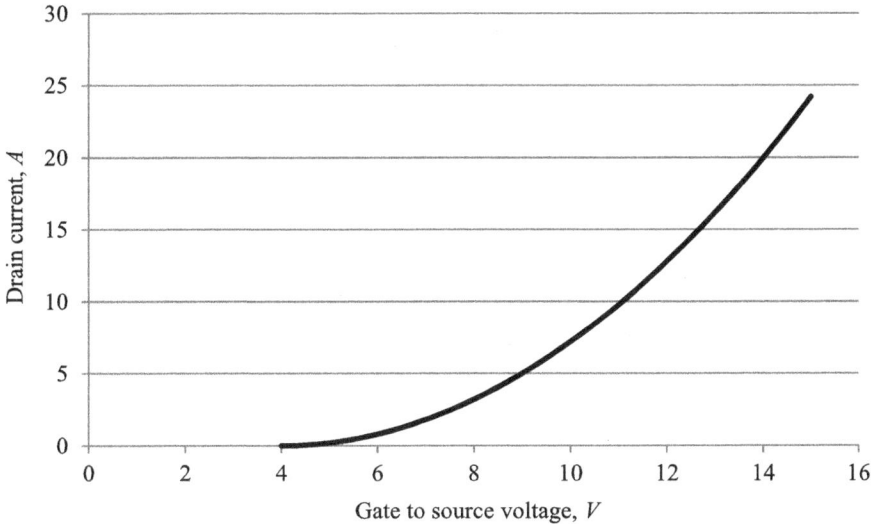

**Figure 4. 15** Drain current against gate-source voltage that is independent of drain-source voltage. The threshold voltage is 4V.

MOSFET conducts at the correct time. Current and voltage information from the power converter can be processed by the controller and combined with time delays so that the correct signals for the transistor gates are generated.

A plot of drain current against gate-source voltage shows that there is no current below a threshold voltage. Above this threshold, the current and voltage follow a square law relationship (Figure 4.15).

Between each of the three terminals is capacitance (Figure 4.16). The gate-source or input capacitance can be several thousands of picofarads. The drain-source or output capacitance is small and typically a few hundreds of picofarads. The gate-drain or reverse transfer capacitance is also a few hundreds of picofarads. A more useful variable is the gate charge (Figure 4.17). Manufacturers will

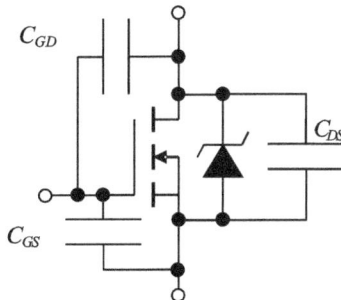

**Figure 4.16** Capacitances between the three terminals of a MOSFET.

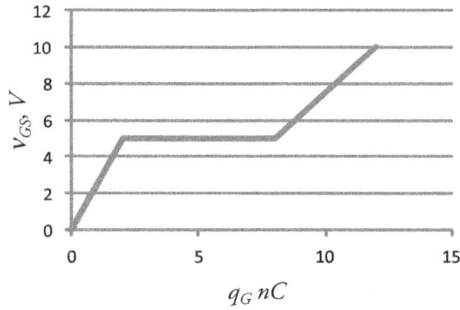

**Figure 4.17**   Gate-source voltage against Gate charge.

state capacitances and charge. These values are useful when calculating the peak current required from a gate drive circuit. For example, a gate capacitance of 100 nC requires 1 mA of current for 1 μs but 5A for 20 ns. A gate circuit for a MOSFET is relatively simple but can require high peak currents for short periods while providing gate charge during turn-on and removing gate charge during turn-off. During periods when a MOSFET is not being switched, negligible gate current is needed due to the very high input resistance of the gate.

A useful parameter for a MOSFET is its drain-source on-state resistance, $R_{DSon}$. When the gate-source voltage is increased above the threshold voltage, the ratio of the drain-source voltage to drain current is approximately constant. This resistance is stated in data sheets for typical gate-source voltages, drain currents, and device temperatures.

## 4.4   Insulated Gate Bipolar Transistor

There are n-channel and p-channel insulated gate bipolar transistors (IGBT) (Figures 4.18 and 4.19). These devices can withstand 1,200V when in their off state and currents of 150A can be conducted continuously in their on state.

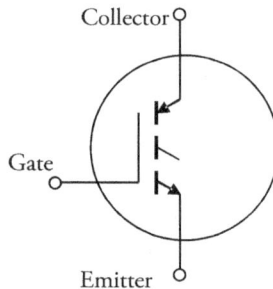

**Figure 4.18**   Circuit symbol for an n-channel IGBT.

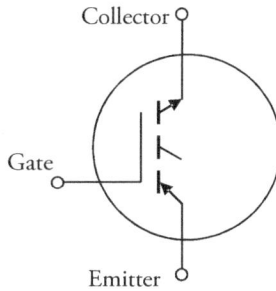

**Figure 4.19**  Circuit symbol for an p-channel IGBT.

Higher current devices are made, but they have a lower voltage. The continuous collector current at 100°C is approximately half that at an ambient temperature of 25°C. They combine the good input characteristics of a MOSFET (high input impedance and simple gate circuit) with the better power handling output characteristics of a BJT (high current and voltage). The typical output characteristic of an IGBT is similar to that of a MOSFET (Figure 4.20) except that the collector to emitter saturation voltage is approximately between 1V and 2V

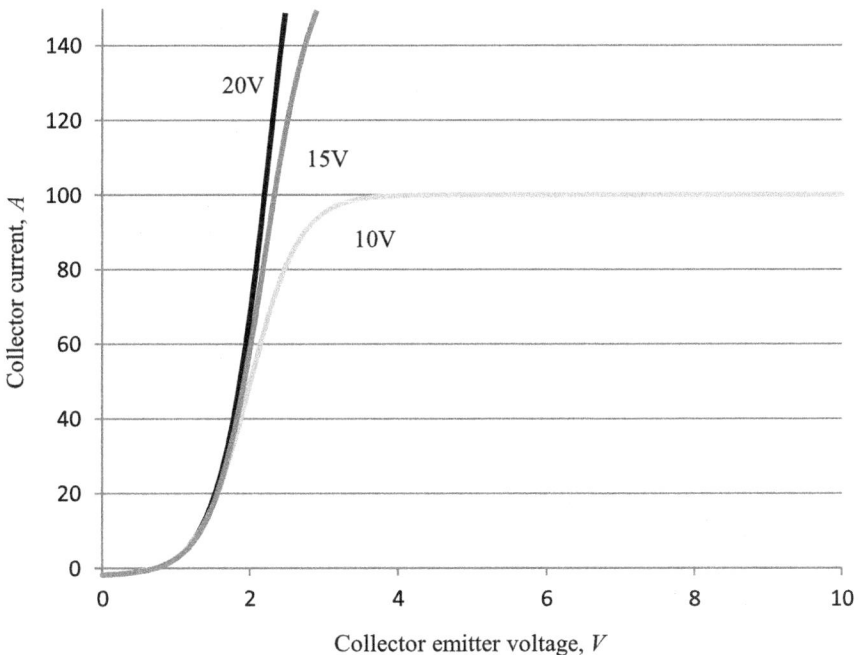

**Figure 4.20**  Output characteristic of an IGBT at three gate emitter voltages for a typical 100A device at 100⁰C.

depending on the rating of the device and the operating temperature. There are also devices that have a diode in parallel with the collector and emitter.

## 4.5   Guide to Power Level and Frequency

At the basic level, the selection of a power semiconductor for a specified circuit requires an idea of the approximate power that the load will put on the output of a converter and the switching speed during its operation. There is no universal device that will fit every situation. A device that has a simple and small input gate circuit with a high off-voltage and high current during conduction does not exist. Generally as the power increases the switching frequency decreases. At one of the spectrum is the thyristor as a low-speed and high-power device, whereas the MOSFET has a low-power capacity but the highest switching speed (Figure 4.21).

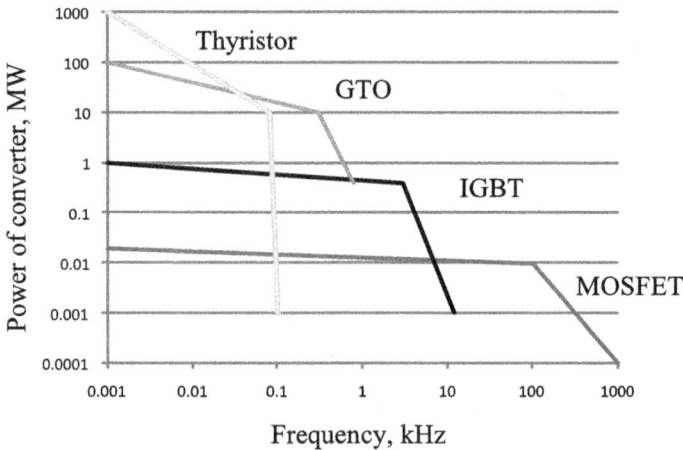

**Figure 4.21**   A guide to the selection of a power device showing the range of power against frequency.

## Selected Bibliography

Davis, R. M., *Power Diode and Thyristor Circuits*, London, UK: Peter Peregrinus, 1976.

Lander, C. W., *Power Electronics*, 3rd ed., New York: McGraw-Hill, 1993.

# 5

# Heating and Cooling

A voltage across a power semiconductor device with a current flowing through it causes heating of the silicon. A manufacturer will specify the practical limit on the device temperature. Heat generated inside the device ($p = vi$) will conduct from the inside through the base (case) to a heat sink and the ambient air or coolant (see Figure 5.1). The temperature inside the device is historically called the junction temperature (pn junction)_$T_j$. Manufacturers specify the maximum temperature at which the device can operate safely. This value is typically 150°C, but depends on the device.

## 5.1 Thermal Resistance

Heat generated inside an enclosure flows out to the external environment where the rate of flow of energy depends on the thermal properties of the material between the source of power and the surroundings (see Figure 5.2).

A thermal resistance, $R_\theta$, has the units of °C W$^{-1}$

$$\Delta T = P\, R_\theta \tag{5.1}$$

$$\Delta T = T_j - T_a \tag{5.2}$$

where $\Delta T$ is the difference in temperature. $P$ is the power generated inside the device, $R_\theta$ is the thermal resistance, $T_j$ is the junction temperature, and $T_a$ is the ambient temperature. For a constant power, the temperature difference is low if the thermal resistance is low. The inside of a power device is then kept cool and more current can flow.

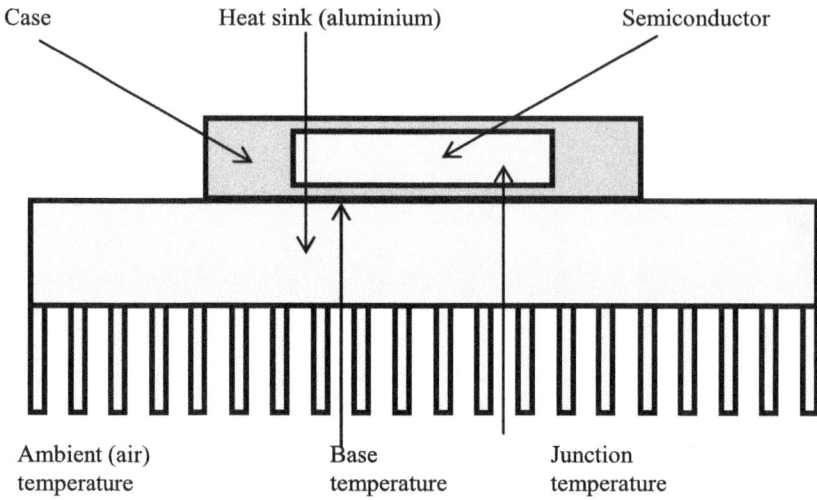

**Figure 5.1**    Cross section through a power semiconductor and heat sink.

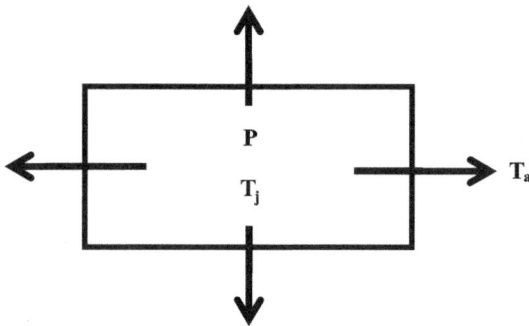

**Figure 5.2**    Power generated inside a boundary causes a temperature difference between the inside and outside of a device.

## 5.2   Equivalent Circuit

Concepts and calculations can be made using an electrical equivalent circuit model (Table 5.1). Hence, Ohm's law can be used for thermal resistance calculations.

The temperature differences across the resistances are given by the following equations:

$$T_c - T_s = R_{\theta(c-s)}\, P \tag{5.3}$$

$$T_j - T_c = R_{\theta(j-c)}\, P \tag{5.4}$$

**Table 5.1**
Duality Between Thermal and Electrical Variables and Parameters

| Thermal | Electrical | Thermal | Electrical |
|---|---|---|---|
| Power | Current | Temperature | Voltage |
| Thermal resistance | Resistance | Thermal capacity (inertia) | Capacitance |

$$T_s - T_a = R_{\theta(s-a)}P \tag{5.5}$$

$$T_j - T_s = R_{\theta(j-s)}P \tag{5.6}$$

where

$$R_{\theta(j-s)} = R_{\theta(j-c)} + R_{\theta(c-s)} \tag{5.7}$$

The thermal capacity, $C_j$, is small and difficult to measure but fluctuates depending on the power and duty cycle. The heat sink capacitance, $C_s$, is large and therefore the temperature, $T_s$, has little fluctuation despite the switching on (temperature rises) and switching off (temperature falls) of a power semiconductor.

## 5.3   Transient Thermal Resistance

As the thermal capacity of power device is difficult to measure, it has been replaced in calculations by a transient thermal resistance (or impedance), $Z_{\theta(j-s)}$. This concept is based on rectangular pulses of power that occur in a repetitive period, $T_p$ with a duty cycle, $\delta$ (Figure 5.3). The transient response of the temperature is a result of the first-order system from the resistance and thermal capacity values (Figure 5.3). The inputs are step changes in power so the power rises and falls exponentially.

$$T = P\left[1 - e^{-t/\tau}\right] \tag{5.8}$$

$$T = P\,e^{-t/\tau} \tag{5.9}$$

The temperature variation is shown in Figure 5.4 for a rectangular power pulse, $P_\tau$. In general a device may have a constant base current, $P_o$. The various

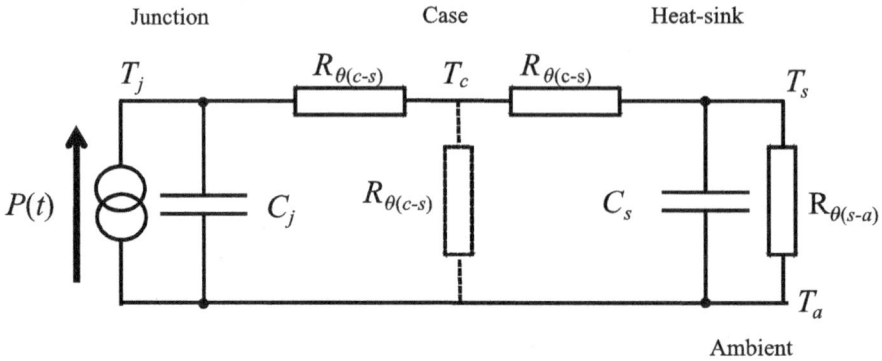

**Figure 5.3**    Electrical model of the thermal effects of heating for a power semiconductor where $P(t)$ is the rate of heat generated watts, $R_{\theta(j-c)}$ is the internal resistance junction case °C W$^{-1}$, $R_{\theta(c-s)}$ is the resistance case of heat sink °C W$^{-1}$, $R_{\theta(s-a)}$ is the resistance heat sink, ambient °C W$^{-1}$, $C_j$ is the thermal inertia of junction J°C$^{-1}$, $C_s$ is the thermal inertia of junction J°C$^{-1}$, $T_j$ is the temperature of junction °C, $T_c$ is the temperature of case °C, and $T_s$ is the temperature of heat sink °C.

temperature differences are shown expressed in the resistance and power terms on the right side of the figure.

The pulse power is calculated from the current waveform and the device forward characteristic. For a p-n junction (2.4) is used. The average junction temperature is calculated from the mean power in the pulse over the period. The

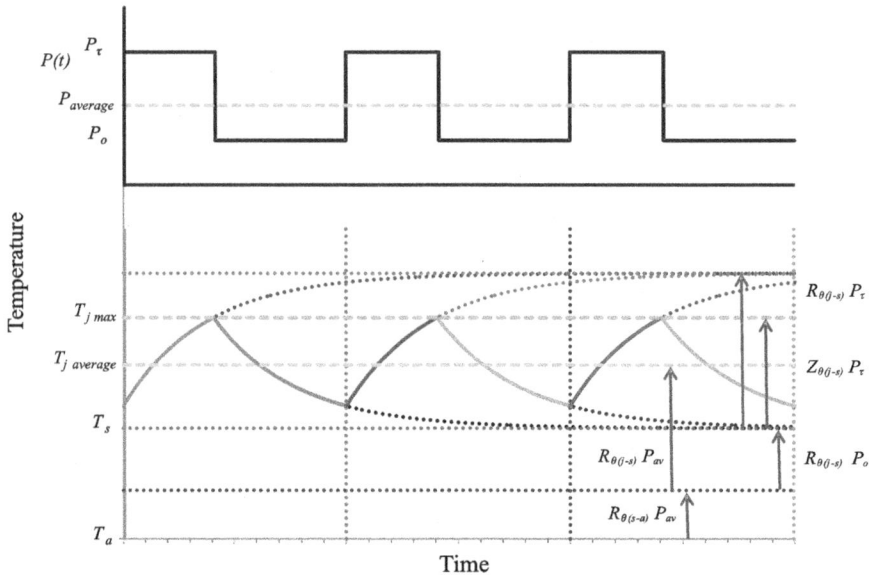

**Figure 5.4**    Junction temperature is shown at the top of the figure as a solid waveform and three rectangular power pulses.

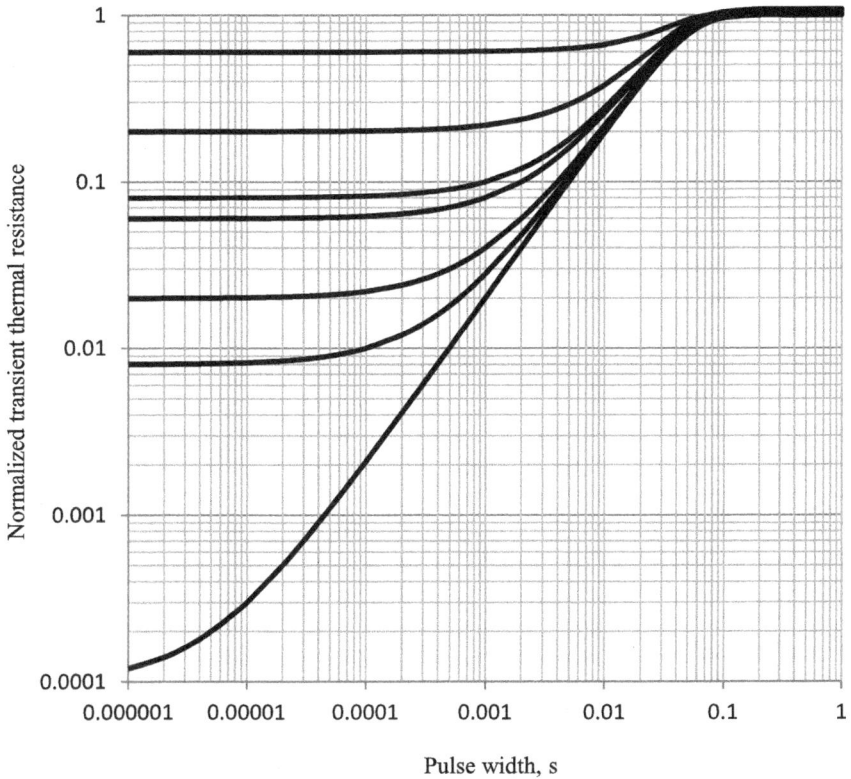

**Figure 5.5** An example of a transient thermal resistance characteristic. The resistance decreases as either the period or duty cycle decrease. The duty cycles for the curves
shown from top to bottom are 0.5, 0.2, 0.1, 0.05, 0.02, and 0.01. The vertical axis shows
the transient thermal resistance normalized relative to the thermal resistance.

transient thermal resistance is found from the data sheet for a particular device
(Figure 5.5). The curves in a plot show values for given duty cycles and periods.
At the end of a power pulse, the temperature reaches its maximum value which
is determined from the pulse power and transient thermal impedance.

## Selected Bibliography

Davis, R. M., *Power Diode and Thyristor Circuits*, London, UK: Peter Peregrinus, 1976.

Lander, C. W., *Power Electronics*, 3rd ed., New York: McGraw-Hill, 1993.

Williams, B. W., *Power Electronics: Devices, Drivers, Applications, and Passive Components*, New
York: Macmillan, 1992.

# 6

# Phase-Controlled Thyristor Converters and Diode Rectifiers

## 6.1 Phase-Controlled AC Thyristor Converter

Connecting two thyristors in parallel creates a device that can control the ac voltage applied to a load (Figure 6.1). It can be made using two separate thyristors connected back to back or a triac. By delaying the time at which the thyristors are turned on, the magnitude of the ac voltage is controlled. One device, $P_1$, controls the voltage and current on positive half-cycles while the second device, $N_1$, controls voltage and current on the negative half-cycles. The voltage waveform applied to the load on the positive half-cycle should be the same shape but inverted on the negative half-cycle. Hence, the delay angle for both half-cycles should be equal; otherwise, a dc component will be present in the load. A simple use of this circuit is in a light dimmer switch for powers in the range of 40W to 250W, where the delay is created by a simple series resistor and capacitor. Varying the resistance will vary the delay and control the light output (Figure 6.2). If inductance is added in series with the resistive load, the current will take time to rise and fall to zero (see Figures 3.14 to 3.20 in Chapter 3 for a single pulse applied to a thyristor). In the extreme case of a purely inductive load, the current will not have time to fall below the holding current of the thyristor before the next gate current is applied to the second thyristor. The result is that there will be a direct current in the load, either positive or negative, depending on which thyristor is fired first.

The RMS output voltage across the load is given by

$$V_{RMS} = \sqrt{\frac{1}{2\pi}\int_{\alpha}^{2\pi} V_m^2 \sin^2\theta\, d\theta} \qquad (6.1)$$

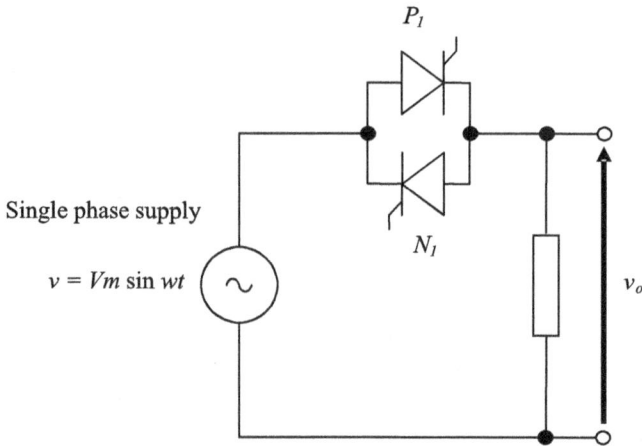

**Figure 6.1** Single-phase-controlled converter with resistive or series-resistive and inductive load.

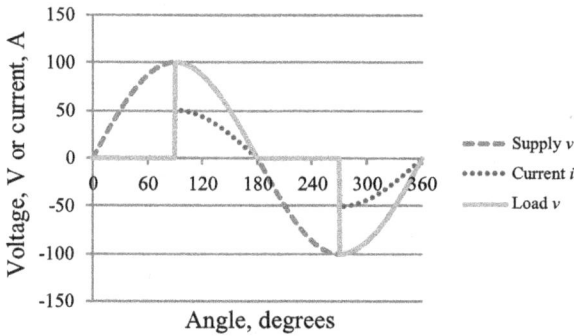

**Figure 6.2** Resistance load and a phase delay of 90°. Supply voltage, current, and load voltage.

where $\alpha$ is the delay angle, $V_m$ is the peak supply voltage, and $\theta = \omega t$.

$$V_{RMS} = \frac{V_m}{\sqrt{2\pi}} \sqrt{\int_\alpha^{2\pi} \left[ \frac{1 - \cos 2\theta}{2} \right] d\theta} \tag{6.2}$$

$$V_{RMS} = \frac{V_m}{2\sqrt{\pi}} \sqrt{\left[ \theta - \frac{1}{2} \sin \theta \right]_\alpha^{2\pi}} \tag{6.3}$$

$$V_{RMS} = \frac{V_m}{\sqrt{2\pi}} \sqrt{\left[ 2\pi - \alpha + \frac{1}{2} \sin \alpha \right]} \tag{6.4}$$

With a single-phase ac supply, a single diode can be used to produce a dc voltage source, either positive or negative. A more useful circuit employs four diodes to rectify both the positive and negative voltage half sine waves. For a three-phase supply, only two extra diodes are needed to make a three-phase rectifier. The output voltage is uncontrolled in the sense that the ac supply fixes the voltage output.

Variation of the output voltage is achieved with phase control of thyristors sets that are naturally commutated by the ac supply voltages. They have a wide range of application in electric drives and power distribution networks. Practical circuits have four thyristors for a single-phase supply and six thyristors for a three-phase supply. The circuits can be classified and analyzed according to the number of events or switching of thyristors that occur in a cycle of the supply voltage. They have a pulse number, $p$. Single-phase converters have a pulse number of two and switching occurs every 180° of the supply cycle. Three-phase converters have a pulse number of six where switching occurs every 60° of the supply cycle. These converters can rectify where energy from an ac source is converted into a dc supply. They can also invert when energy from a dc source is converted into an ac supply.

## 6.2  Single-Phase Converter

This converter consists of four diodes (Figure 6.3). In Figure 6.4, on the positive half-cycle, current flows from the supply through one of the top diodes ($P_1$) and then the load and is returned to the supply via one of the lower diodes ($N_1$). The other two diodes ($P_2$ and $N_2$) take over the current during the negative half-cycle (Figure 6.5). The circuit converts an ac voltage and current supply into a dc voltage and current supply (Figure 6.6). This process is called rectification.

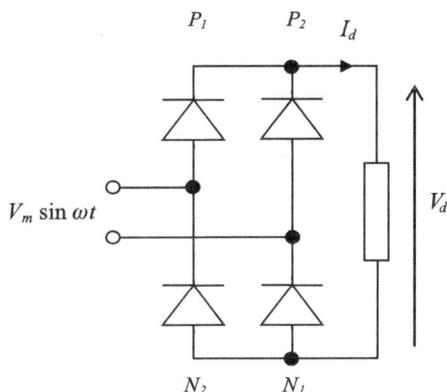

**Figure 6.3**  Single-phase diode bridge.

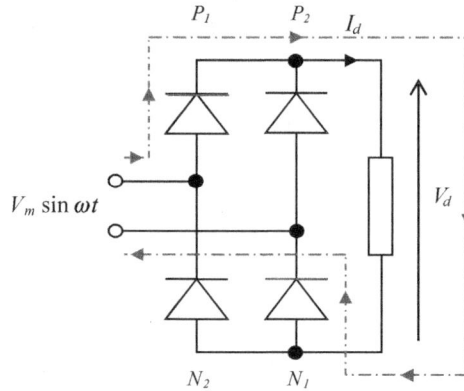

**Figure 6.4**    Positive half-cycle current flow.

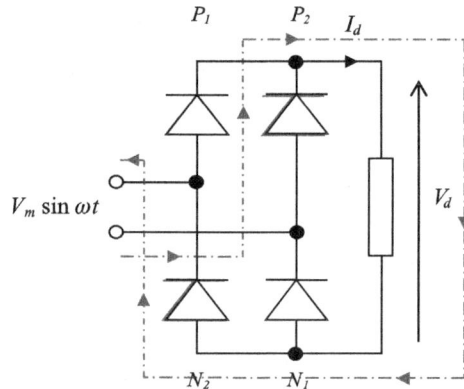

**Figure 6.5**    Negative half-cycle current flow.

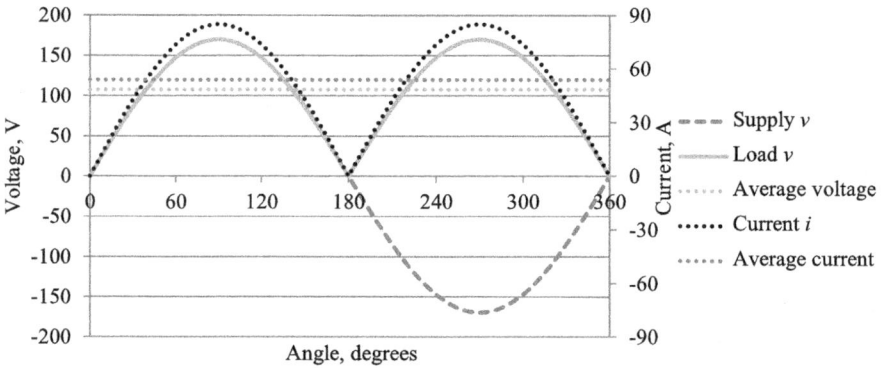

**Figure 6.6**    Rectified 120-V single-phase supply with a resistive load of 2Ω. Note that the vertical axis for the current is shown on the right side.

The voltage in Figure 6.6 is given by

$$v = V_m \sin\theta \tag{6.5}$$

The rectified and averaged voltage is

$$V_{av} = \frac{1}{2\pi} \int_0^{2\pi} |V_m \sin\theta| \, d\theta \tag{6.6}$$

The integration can be taken over a half of a supply cycle since the waveform repeats every 180°.

$$V_{av} = \frac{V_m}{\pi} [-\cos\theta]_0^\pi \tag{6.7}$$

$$V_{av} = \frac{2V_m}{\pi} \tag{6.8}$$

and the average current is simply

$$I_{av} = \frac{2V_m}{\pi R} \tag{6.9}$$

Replacing the diodes with thyristors allows for phase control where the output voltage and hence power from the input to the output can be varied by delaying the time (angle) at which the thyristors are triggered. The thyristors shown in Figure 6.7 are triggered in pairs (1 and 2 together, 3 and 4 together) in the same natural pattern as if they were diodes. The output voltage appears as parts of a half sine wave (Figure 6.8).

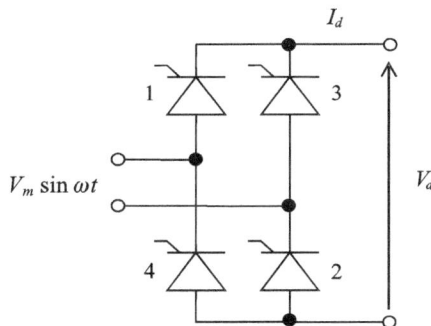

**Figure 6.7**   Single-phase-controlled thyristor converter.

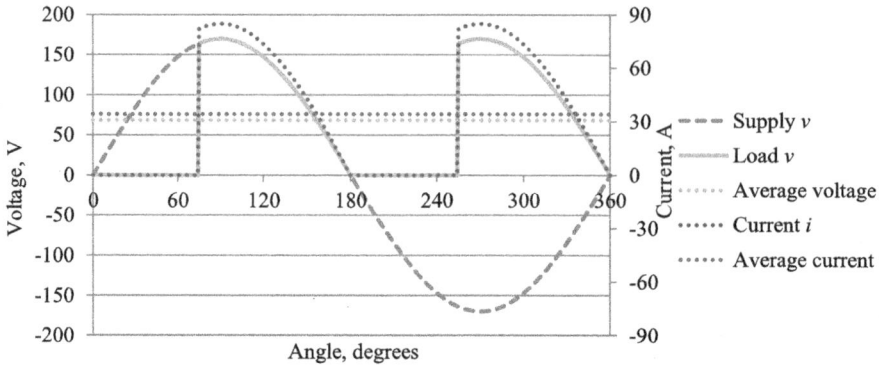

**Figure 6.8**   A delay angle of 75° for rectified 120-V single-phase supply with a resistive load of 2Ω.

The average output voltage, $V_{av}$, as a function of the delay angle, $\alpha$, is

$$V_{av} = \int_{\alpha}^{\pi} V_m \sin\theta \; d\theta \quad where \;\; \theta = \omega t \tag{6.10}$$

$$V_{av} = \frac{V_m}{\pi}(1 + \cos\alpha) \tag{6.11}$$

These equations show that there is a nonlinear relationship between the delay angle, $\alpha$, and the average output voltage. The maximum voltage occurs when the delay angle is zero. The sense of the control of average voltage is counterintuitive (i.e., at start-up), a low voltage is achieved with a delay angle near 180°.

## 6.3   Three-Phase Converter

To extend a single-phase rectifier (Figure 6.3) to a three-phase rectifier only requires an extra two devices (Figure 6.9).

Each diode conducts for one-third of the cycle (120° or $2\pi/3$ radians). If a diode in the switching sequence is conducting current and is a member of the top three set (labeled $P$ in Figure 6.9), then the next one to conduct will be in the bottom set (labeled $N$). The conduction sequence alternates between diodes in the top set and ones in the bottom set (Figure 6.10). The output voltage has six pulses of 60° corresponding to the peak top part of the voltage waveforms from 60° to 120° (Figure 6.11). For a purely resistive load and no circuit inductance, the six-step waveform from 30° to 90° is the part of a sinusoidal function that ranges from 60° to 120°. The next 60° from 90° to 150° also has the same top

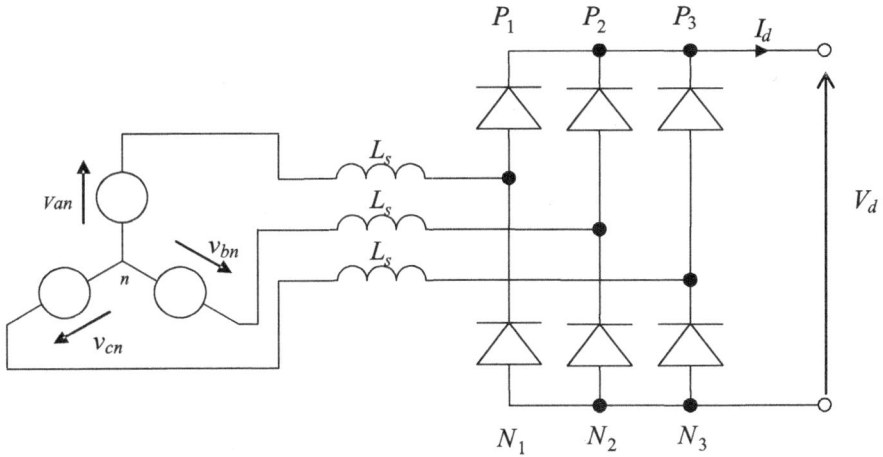

**Figure 6.9** Three-phase diode bridge rectifier with leakage inductances, $L_s$.

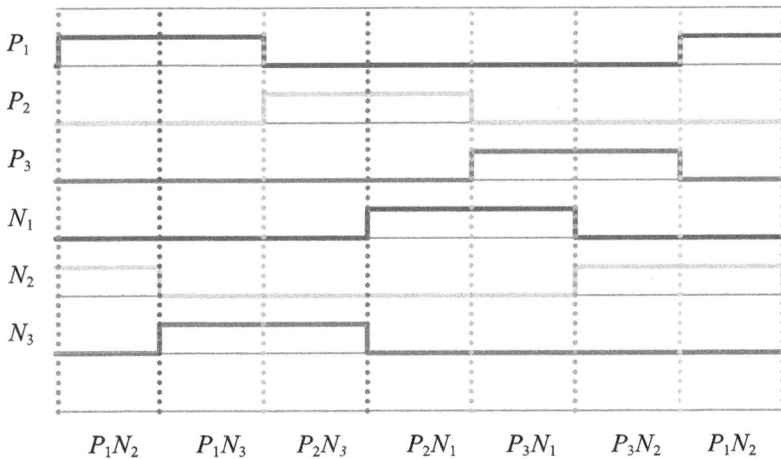

**Figure 6.10** Switching sequence for a three-phase diode rectifier, showing two diodes conducting at any instant of time, one from the positive set and one from the negative set.

part of a sine wave. Similarly, from 210° to 330°, the negative 120° block of current has two 60° components (Figure 6.12). The presence of inductance in the circuit, either from leakage in the supply phases ($L_s$) or in the load, results in the smoothing of the 60° ripple. Because each diode conducts for 120°, the phase currents have the sequence divided into 60° events that are zero current for 60°, 120° positive current, zero current for 60°, and 120° negative current. This waveform shape is referred to as a quasi-square or six step. At any instant, there is a line that has zero current, except for the short durations when an outgoing

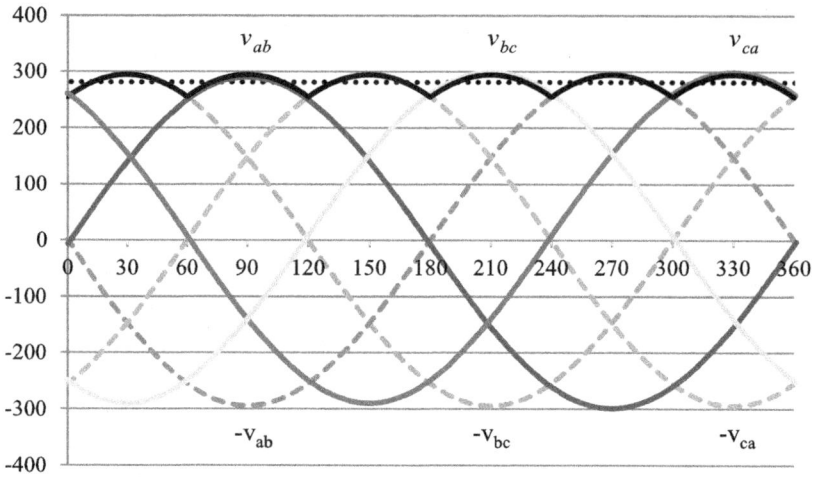

**Figure 6.11**   Three-phase waveforms for a diode converter. The phase shifted 120° line supply waveforms are shown with their inverse. The average output voltage is determined from the peak 60° of each voltage waveform (from 60° to 120°).

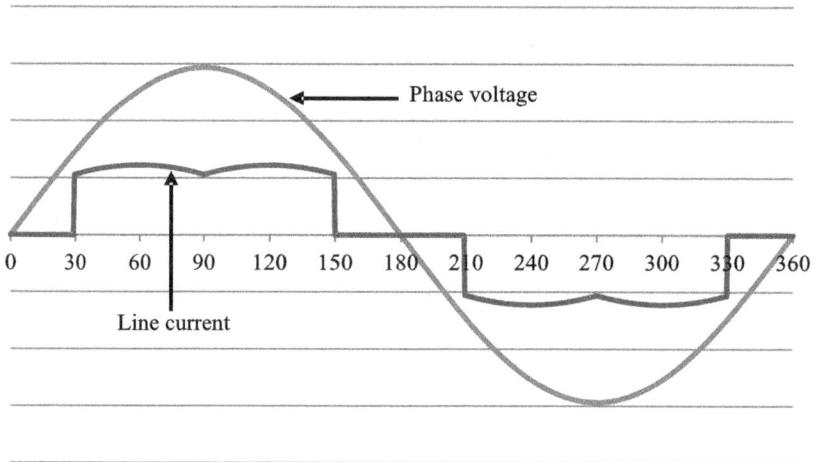

**Figure 6.12**   Line currents with a resistive load.

diode is ceasing to conduct and an incoming diode is being forward biased and taking over the load current. When there is some circuit inductance either in the supply or the load, the blocks of 120° are flat (Figure 6.13). Very little inductance is required in a circuit to achieve this effect.

Changing the six diodes to thyristors allows for phase control of a converter where, by adjusting a delay angle controls, the average voltage at the output

(Figure 6.14). A circuit symbol for these six thyristors is shown in Figure 6.15. Power also can be transmitted from the dc side to the ac side, which is called inversion. The output voltage is raised or lowered by altering a delay angle, $\alpha$, and is the same concept as the control of the single-phase circuit. However, the reference point in time (angle) at which thyristors are triggered is at 60° ($\prod/3$

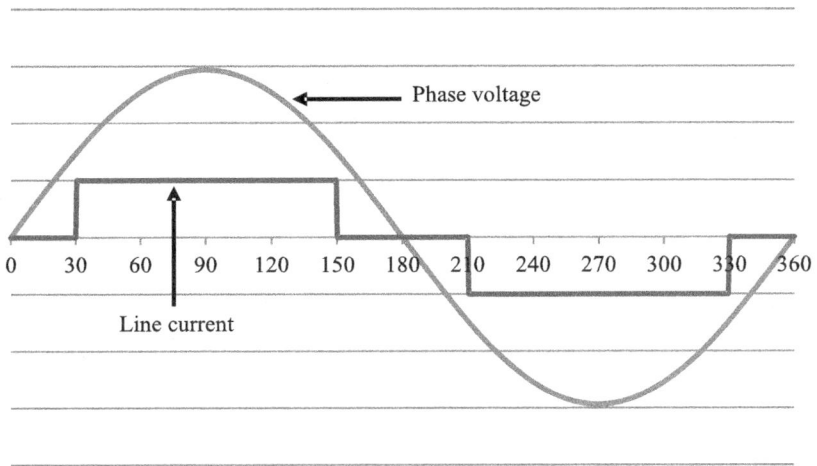

**Figure 6.13**   Line currents with an inductive load.

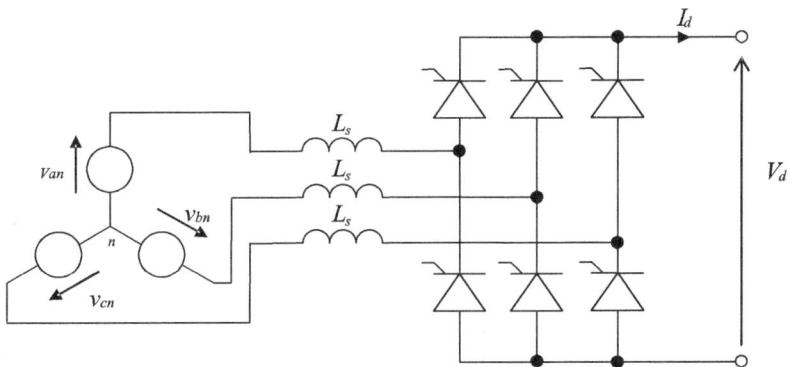

**Figure 6.14**   Three-phase thyristor converter.

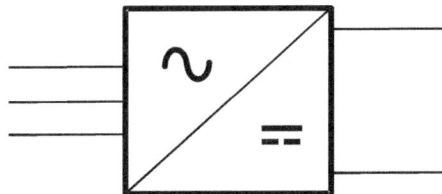

**Figure 6.15**   Circuit symbol for a three-phase thyristor converter.

radians) from the line voltage zero crossing points. These points correspond to the line voltages at 60° intervals (Figure 6.16).

The delay angle ranges from 0° to 180° as shown in Figures 6.11 and 6.16 to 6.21. At 0° the average output voltage is at maximum positive voltage. At 90° the average output voltage is zero. At 180° the voltage is fully negative.

For a p-pulse converter, the average output voltage is calculated by integrating the supply waveform over the pulse period of $2\pi/p$ (Figure 6.22).

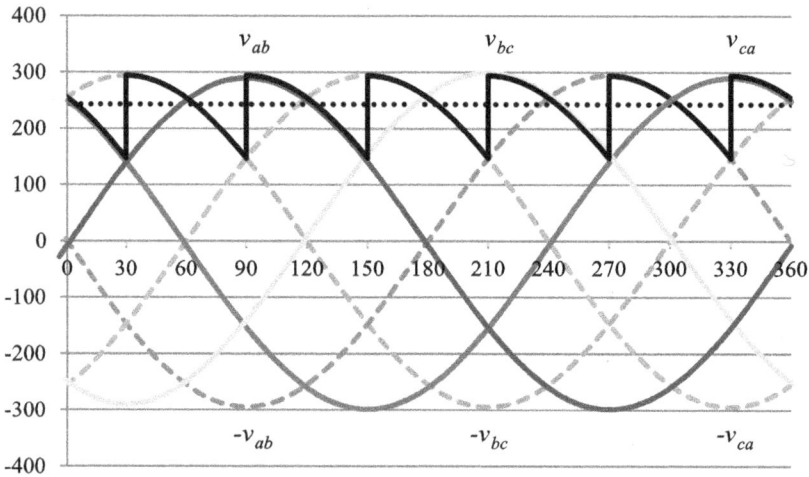

**Figure 6.16**   A 30° delay angle. Output waveforms for a three-phase converter.

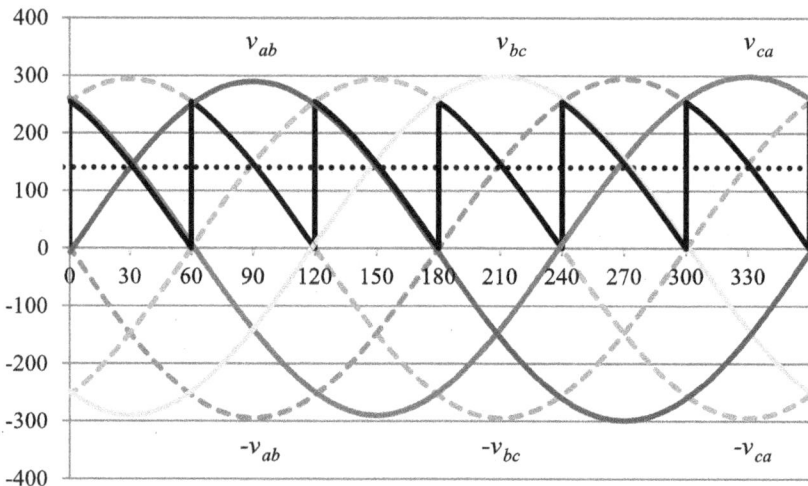

**Figure 6.17**   A 60° delay angle. Output waveforms for a three-phase converter.

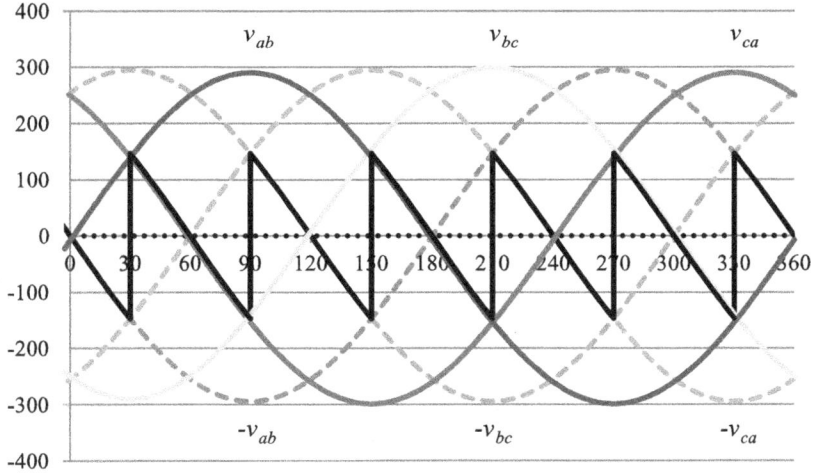

**Figure 6.18**  A 90° delay angle. Output waveforms for a three-phase converter.

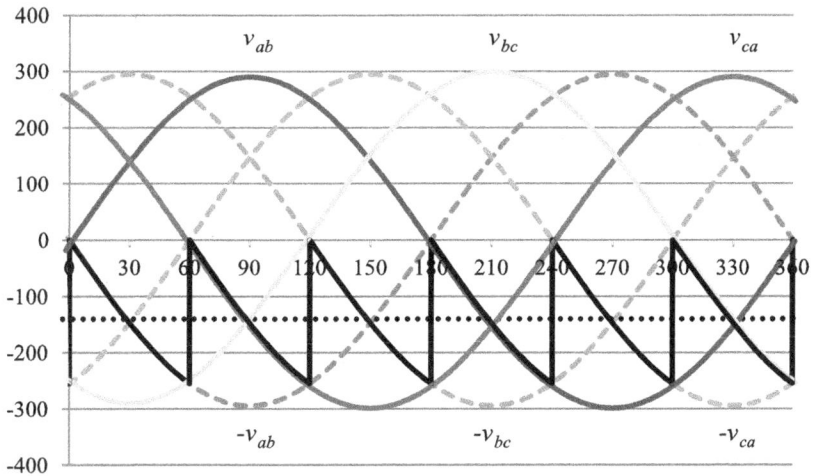

**Figure 6.19**  A 120° delay angle. Output waveforms for a three-phase converter.

$$V_d = \frac{p}{2\pi} \int_{\frac{-\pi}{p}+\alpha}^{\frac{\pi}{p}+\alpha} V_m \cos\theta \, d\theta \qquad (6.12)$$

$$V_d = \frac{p V_m}{2\pi} \left[ \sin\left(\frac{\pi}{p} + \alpha\right) - \sin\left(\frac{-\pi}{p} + \alpha\right) \right] \qquad (6.13)$$

$$V_d = p \frac{V_m}{\pi} \sin\frac{\pi}{p} \cos\alpha \qquad (6.14)$$

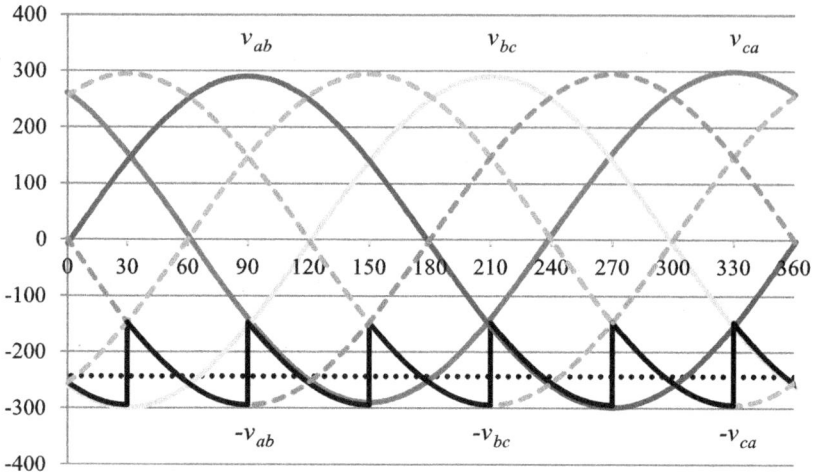

**Figure 6.20**   A 150° delay angle. Output waveforms for a three-phase converter.

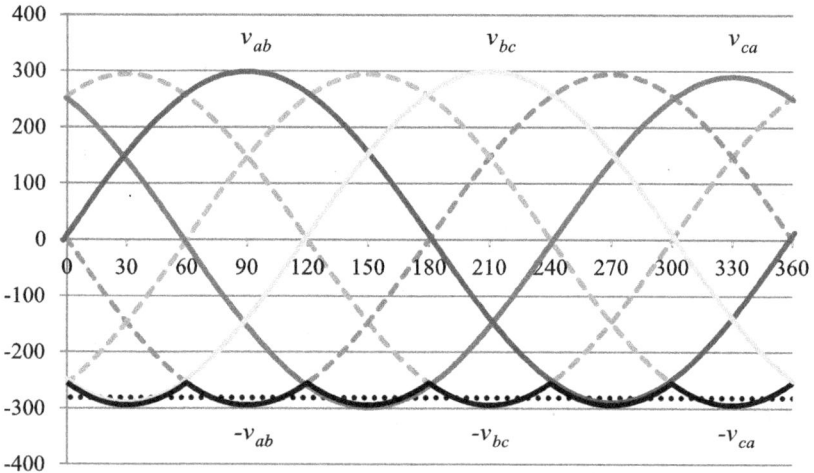

**Figure 6.21**   A 180° delay angle. Output waveforms for a three-phase converter.

For a single-phase converter, the pulse number is two and the average output voltage is

$$V_d = \frac{2V_m}{\pi}\cos\alpha \qquad (6.15)$$

For a three-phase converter, the pulse number is two and the average output voltage is

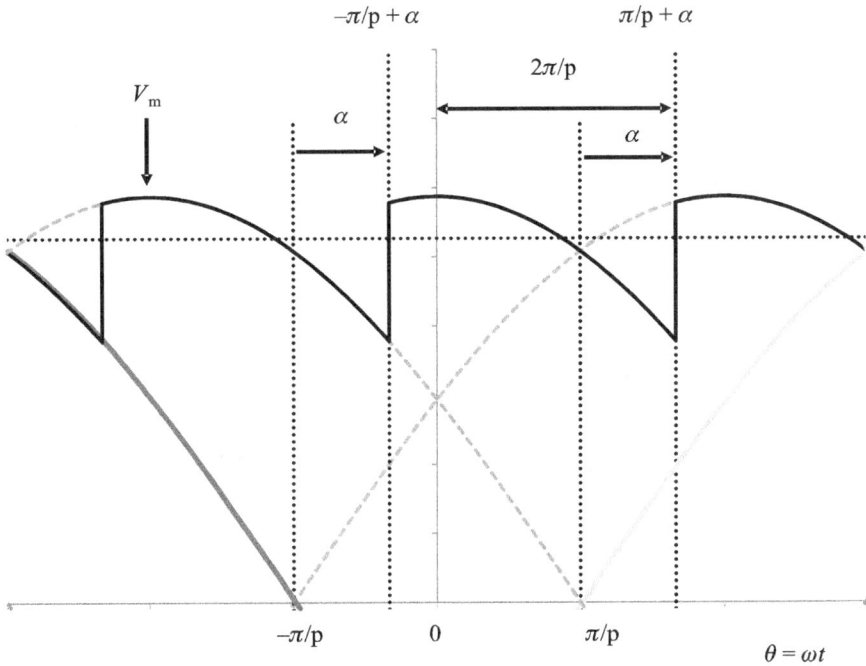

**Figure 6.22** Waveform for a general p-pulse converter showing the delay angle, $\alpha$, and the pulse period, $2\Pi/p$.

$$V_d = \frac{3V_m}{\pi} \cos \alpha \qquad (6.16)$$

Notice that the sense of operation is counterintuitive for positive voltages. When the delay angle is zero, then the dc voltage is at its maximum, and when it is 90°, the voltage is zero.

## 6.4 Overlap

Triggering the next thyristor in the sequence ideally should occur smoothly and instantaneously, but a thyristor has a finite turn on time. Also, the presence of phase inductance either from cables supplying the converter or from transformer windings results in a phenomenon called overlap. Triggering the next thyristor results in two devices with current, and, hence, there is a short circuit between the two lines. The angle (time) over which this process occurs depends on the current, circuit inductance and supply frequency. The load current, $I_d$, is constant during this process. Consider a transfer of current from the "a" line to the "b" line in Figure 6.23.

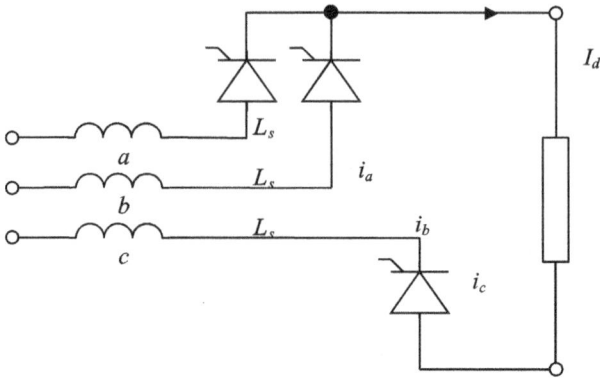

**Figure 6.23**  Overlap where there is a temporary short circuit across the *a* to *b* lines.

$$I_d = i_a + i_b \tag{6.17}$$

During the transfer of current from one thyristor to the next in the sequence the voltage at the cathodes of the two thyristors is halfway between that of the two line voltages (Figure 6.24). Assuming that the passive components of each line are equal, that is, the phase inductances are the same then the loss of voltage due to overlap can be calculated. The current rises from zero to $I_d$ in one thyristor while falling from $I_d$ in the second thyristor.

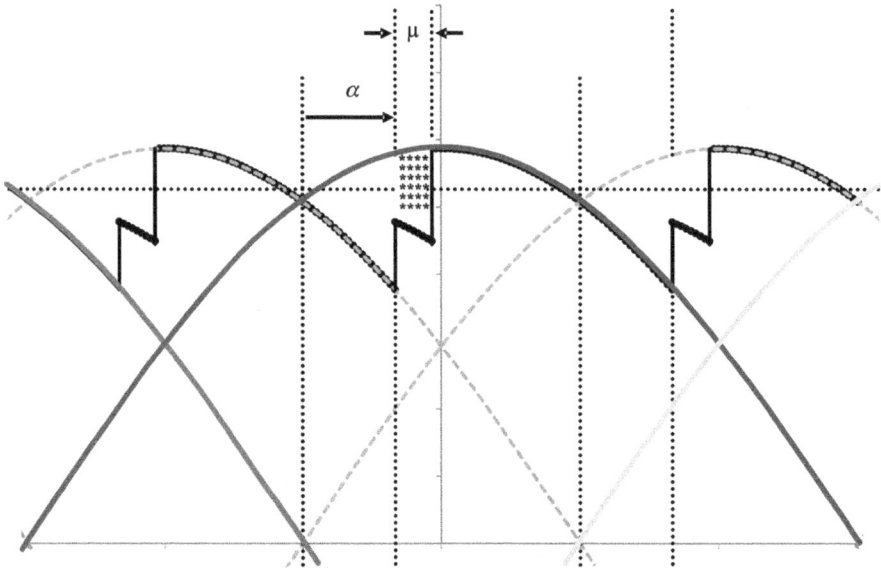

**Figure 6.24**  Overlap angle, $\mu$, and reduction in voltage. The starred area is lost in this process.

The reduction in voltage area, $A$, during the commutation from one thyristor to the next is

$$A = \int_{\alpha}^{\alpha+\mu} v \, d\theta \qquad (6.18)$$

However,

$$\theta = \omega t \qquad d\theta = \omega \, dt \qquad (6.19)$$

$$A = \omega \int_{0}^{T_p} v \, dt \qquad (6.20)$$

The angle of overlap is determined by substituting the equation for an inductor (6.21) into the area equation (6.20).

$$v = L \frac{di}{dt} \qquad (6.21)$$

$$A = \omega \, L_s \int_{0}^{I_d} di \qquad (6.22)$$

$$A = \omega \, L_s I_d \qquad (6.23)$$

Averaging the reduction in area

$$V_{av} = \frac{\omega L_s I_d}{\pi/3} = \frac{3\omega L_s I_d}{\pi} \qquad (6.24)$$

note that this is a reduction in voltage, as the energy in one inductance is transferred to the second inductance. It does not result in a loss of power.

With this expression for overlap, the output voltage from a three-phase converter is

$$V_d = \frac{3V_m}{\pi} \cos\alpha - \frac{3\omega L_s I_d}{\pi} \qquad (6.25)$$

This is an important equation for calculations and control of a three-phase thyristor converter. If the load current is zero, then there is no reduction in voltage [the negative term in (6.25) is zero]. Also, as the load increases the reduction is linearly related to the current.

For a p-pulse converter the dc voltage is

$$V_d = \frac{pV_m}{\pi} \sin\frac{\pi}{p} \cos\alpha - \frac{p\omega L_s I_d}{2\pi} \tag{6.26}$$

The equation for average voltage ignores the voltages across the thyristors. At any instant, there will be two thyristors that are conducting current. Typically this voltage (less than 3V) can be ignored when compared to the voltages in a power system.

If the delay angle is zero, then the line currents and voltages are in phase. As the angle increases, they become out of phase, as shown in Figures 6.25 to 6.31.

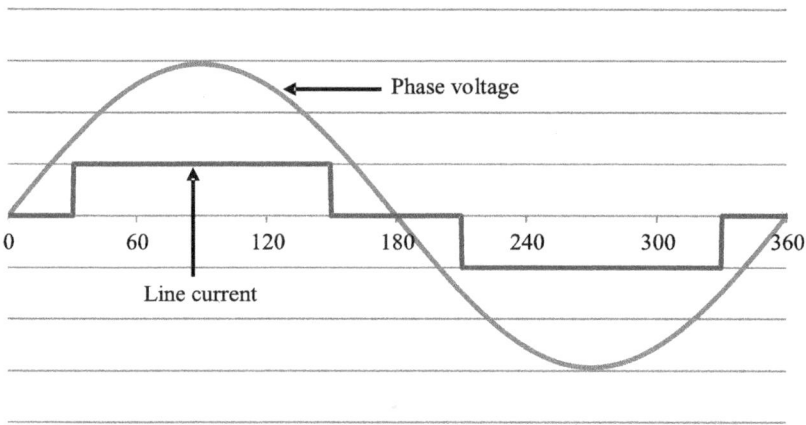

**Figure 6.25**   Zero phase between current and voltage (rectifying).

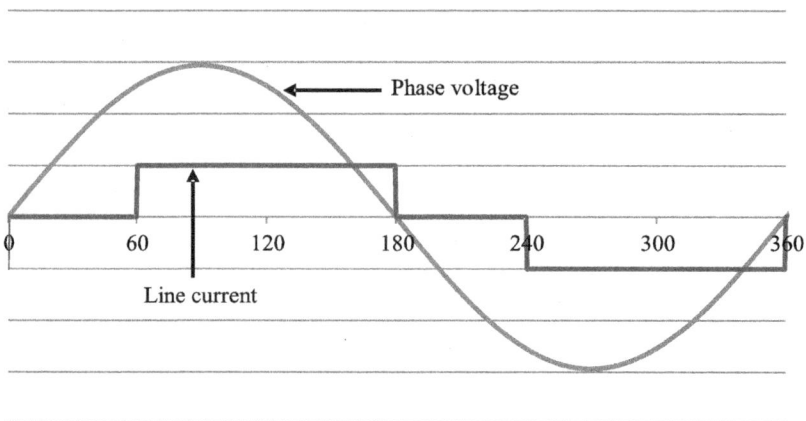

**Figure 6.26**   A 30° phase between current and voltage.

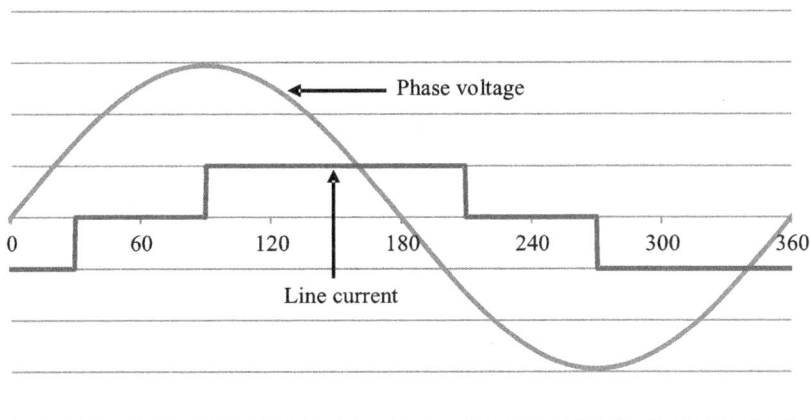

**Figure 6.27**  A 60° phase between current and voltage.

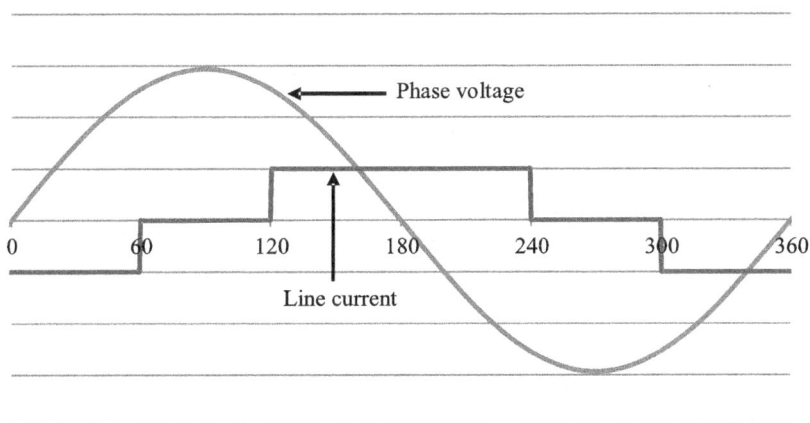

**Figure 6.28**  A 90° phase between current and voltage.

## 6.5  Inversion

If the output of a converter is connected to a source of power, then it can transfer energy to the input supply. To achieve this, transfer requires the delay angle to be advanced from 90° towards 180°. In practice, the chosen angle is not 180°, corresponding to the full negative voltage, but is less than this to account for overlap. A commutation failure will occur if the delay angle is set at or close to 180° with a corresponding surge in load current. The delay angle can also be expressed as an extinction angle, $\beta$.

$$\beta = 180 - \alpha \qquad (6.27)$$

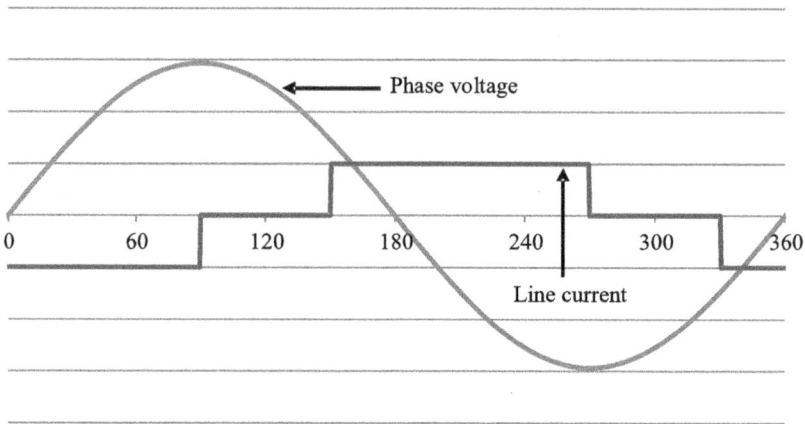

**Figure 6.29**   A 120° phase between current and voltage.

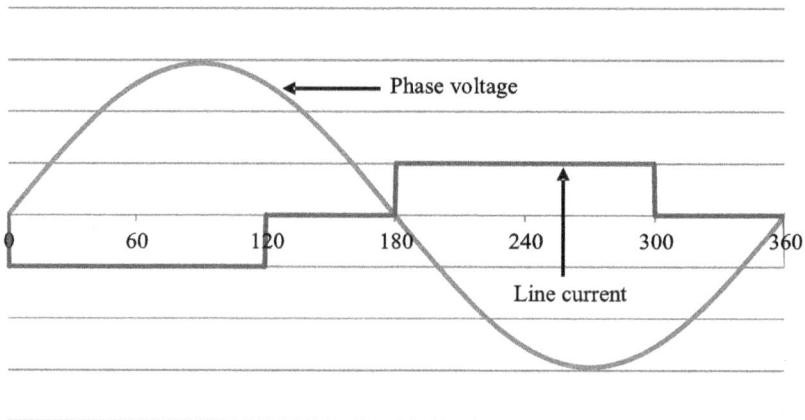

**Figure 6.30**   A 150° phase between current and voltage.

In Figures 6.19 to 6.21, the average voltage (shown as dashed lines in the figures) is negative while the current is positive. Because power is the product of voltage and current, the product is negative. This product does not show that power is created but indicates that the direction of power has changed; power is transferred from the output to the input alternating supply. At full inversion and with no overlap in Figure 6.31 (an angle of 180°), the supply current is inverted with respect to the supply voltage.

## 6.6   Power Systems

A three-phase and fully controlled converter is a very useful building block for many applications with a wide range of power. In power systems energy is

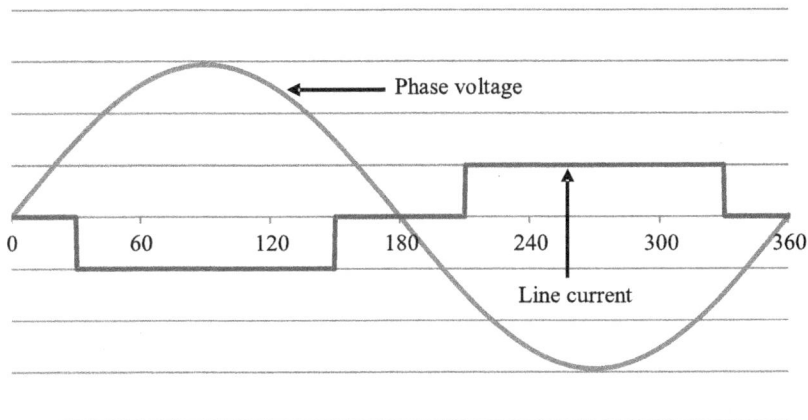

**Figure 6.31**  A 180° phase between current and voltage.

transferred from one country to another where they are separated by the sea using pairs of cables with direct current (Figure 6.32). The advantages of using dc transmission are shown in Table 6.1.

The electrical specification at one end can be different at the other end. For example, a 50-Hz and 415-V supply can be connected via a dc cable to a 60-Hz and 208-V supply. There has to be communication between the two sides as one will be transmitting power and the other will be receiving power. An example of this system is the connection between Sellindge in England and the Les Mandarins station in France. It is a 2,000-MW system with 45 km of cable operating at 270-kV dc.

Power is transmitted from system A to system B by operating the first converter as a rectifier with delay angles between 90° and 0°. Converter B is then operating as an inverter with delay angles between 90° and less than 180°. As the direct current and power are increased, the voltages across the reactive

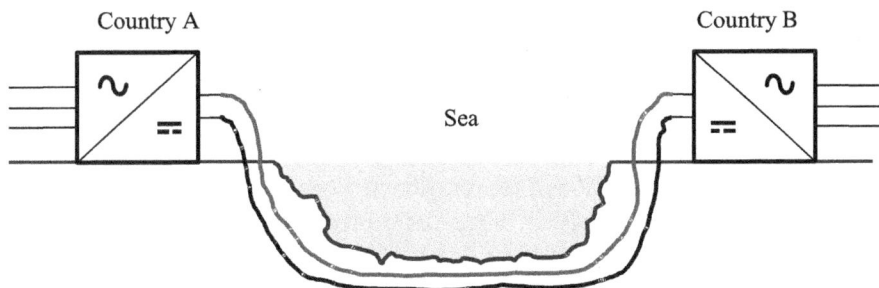

**Figure 6.32**  Conceptual drawing showing two power converters connecting power supplies to two countries.

**Table 6.1**
Advantages of AC-to-DC and DC-to-AC Power Transmission

| |
|---|
| Operating frequencies do not have to be equal. |
| Line voltages do not have to be equal. |
| Clocks can be different. |
| Power can flow in either direction. |
| They have two, not three, cables. |

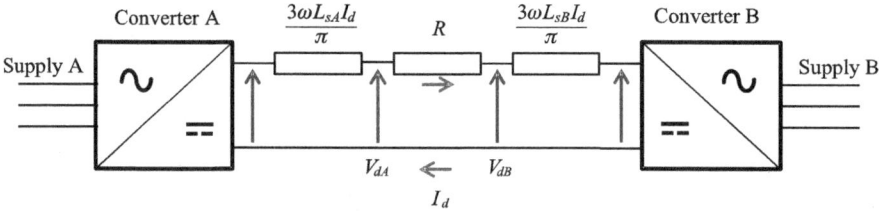

**Figure 6.33**  Block diagram of a dc power transmission system showing power transferred from system A to system B. For transmission in the opposite direction the current is reversed. The resistance, $R$, is the combined resistance of the two cables.

components at either end increase as does that across the cable total resistance. Power is dissipated in the cable resistance and not in the two reactive components. An estimate of efficiency, but not taking into account other smaller losses such as the power dissipated in the thyristors, is

$$\eta = \frac{100(V_{dA}I_d - I_d^2 R)}{V_{dA}I_d} \tag{6.28}$$

The components of the system are shown in Figure 6.33.

## 6.7  DC Motor Drive

A phase-controlled thyristor converter can convert electrical energy into mechanical energy when it is connected to a dc motor (Figure 6.34). By reversing the direction of torque in the motor shaft, mechanical energy can be converted to electrical energy, which is regeneration using delay angles greater than 90°. Connection of two converters across the armature of a dc machine provides for positive and negative voltage as well as positive and negative current (Figure 6.35). Power can be converted to or from the electrical system with the

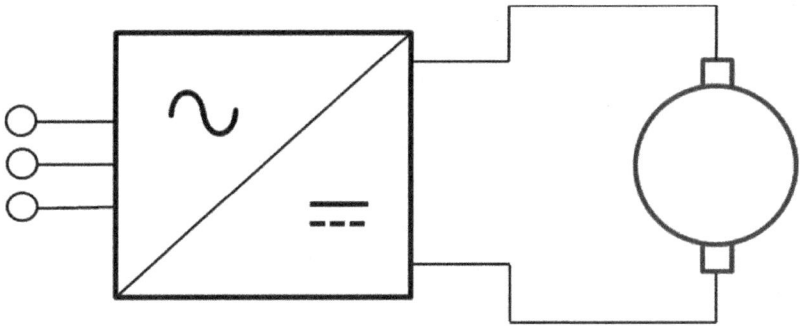

**Figure 6.34**  Circuit diagram of a three-phase converter powering a dc motor.

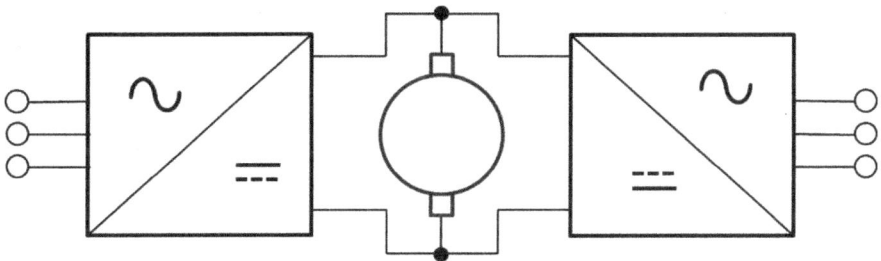

**Figure 6.35**  Four quadrant operation of a dc using two thyristor converters to allow for either polarity of voltage and current.

motor shaft rotating in either a clockwise or counterclockwise sense. Reversing the voltage across the armature requires switching between the two converters. One converter will need to be completely turned off (zero current) before the second converter is activated. A dead band is therefore essential for the reliable operation of the two converters in this topology.

Converting mechanical power into electrical power and vice versa can be illustrated by considering the interchange of energy in an electric vehicle. Accelerating the vehicle in a forward direction is achieved by taking energy from the battery and converting it into mechanical energy. Similarly, the vehicle can be accelerated in the reverse direction using energy from the battery. If the vehicle is traveling forward, it can be decelerated by reversing the direction of torque in the motor shaft. Mechanical energy (kinetic energy) is then converted into electrical energy (kinetic energy recovery). Similarly, while motoring in reverse, mechanical energy can be converted to electrical energy. For any electric drive, the interchange between electrical and mechanical energy can be divided into single-, two-, and four-quadrant operations (Figure 6.36).

Positive
torque

Reverse
regenerating

Forward
motoring

Negative

Positive

Rotational
speed

Reverse
motoring

Forward
regenerating

Negative

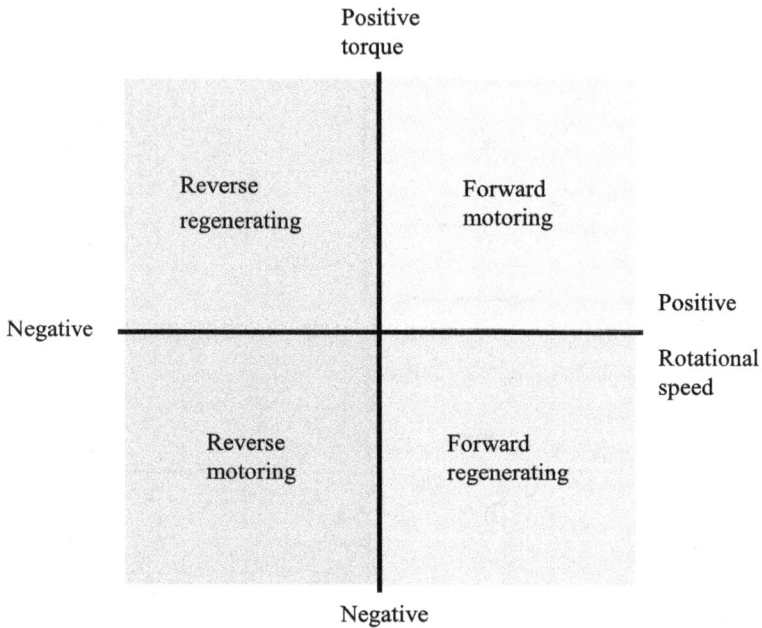

**Figure 6.36**   Diagram of four-quadrant operation for an electric drive.

## Selected Bibliography

Davis, R. M., *Power Diode and Thyristor Circuits*, London, UK: Peter Peregrinus, 1976.

Dewan, S. B., G. R. Slemon, and A. Straughen, *Power Semiconductor Drives*, New York: John Wiley & Sons, 1984.

Koss, A., *A Basic Guide to Power Electronics*, New York: John Wiley & Sons, 1984.

Lander, C. W., *Power Electronics*, 3rd ed., New York: McGraw-Hill, 1993.

Williams, B. W., *Power Electronics: Devices, Drivers, Applications, and Passive Components*, New York: Macmillan, 1992.

# 7

# Cycloconverter

The cycloconverter is a device that converts ac voltages to ac voltages at a frequency that is lower than the supply frequency. It is mainly a high-power converter for variable speed drives in high-torque and low-speed applications where there is large load inertia.

## 7.1 Single-Phase Cycloconverter

A single-phase thyristor converter can output half-sinusoidal waveforms of voltage with varying magnitude depending on the chosen delay angle. A second converter that is connected to the first but with opposite polarity can form negative half-sinusoidal waveforms (Figure 7.1). If the first positive converter outputs two identical half-sinusoidal waveforms and the second negative converter outputs two negative half-sinusoidal waveforms, then the resultant waveform is at half the supply frequency (Figure 7.2). Similarly, three or more positive pulses followed by an equal number of negative pulses produce a lower-frequency output voltage at one-third of the supply frequency. A 50-Hz supply with five positive half-sinusoidal waveforms and five negative half-sinusoidal waveforms forms a 10-Hz output. This technique results in a distorted output which in appearance is not close to the desired pure sinusoidal waveform at 10 Hz. An improvement is to select a large delay angle for the first and fifth half-cycles when the voltage should be low (Figure 7.3). Reducing the delay angle for the second and fourth half-cycles increases the voltage at these times. The peak voltage should occur for the third half-cycle corresponding to the peak of the sinusoidal waveform at 10 Hz. The same delay angles are then used for the negative half-cycles.

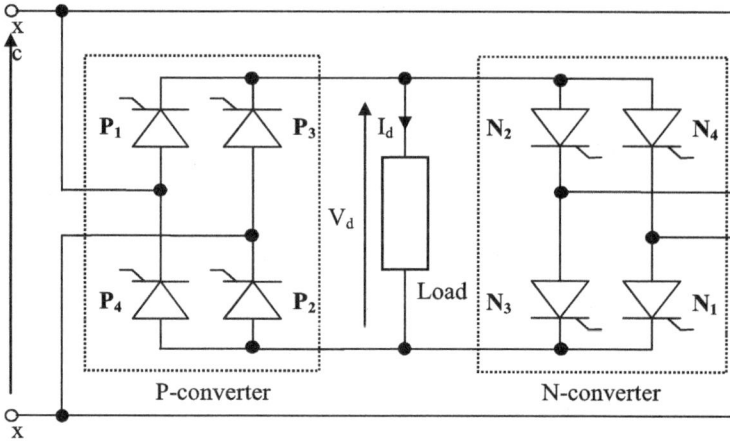

**Figure 7.1**  Single-phase cycloconverter.

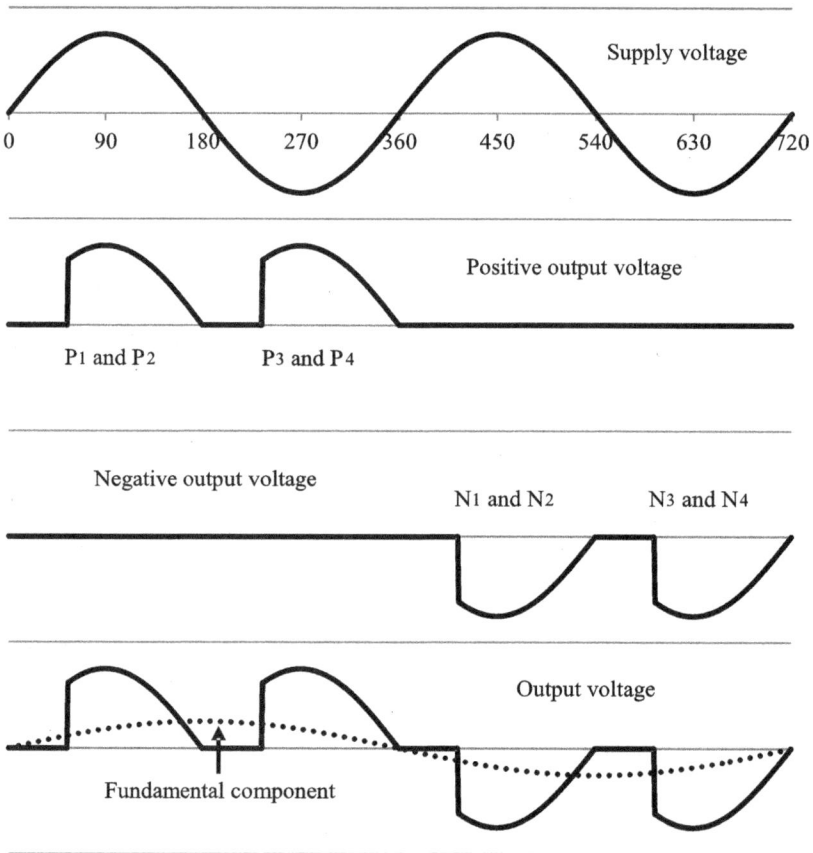

**Figure 7.2**  Cycloconverter with an output voltage whose frequency is half that of the supply.

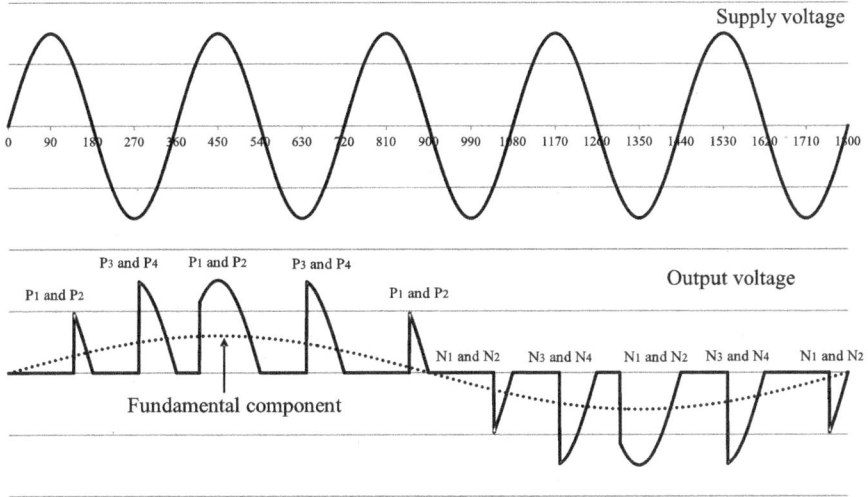

**Figure 7.3** Cycloconverter with an output voltage whose frequency is one-fifth that of the supply.

The average voltage over a supply half-cycle with a delay angle of $\alpha$ is

$$V_{av} = \frac{1}{\pi} \int_{\alpha}^{\pi} V_m \sin\theta \, d\theta \qquad (7.1)$$

$$V_{av} = \frac{1}{\pi}[-V_m \cos\theta]_{\alpha}^{\pi} \qquad (7.2)$$

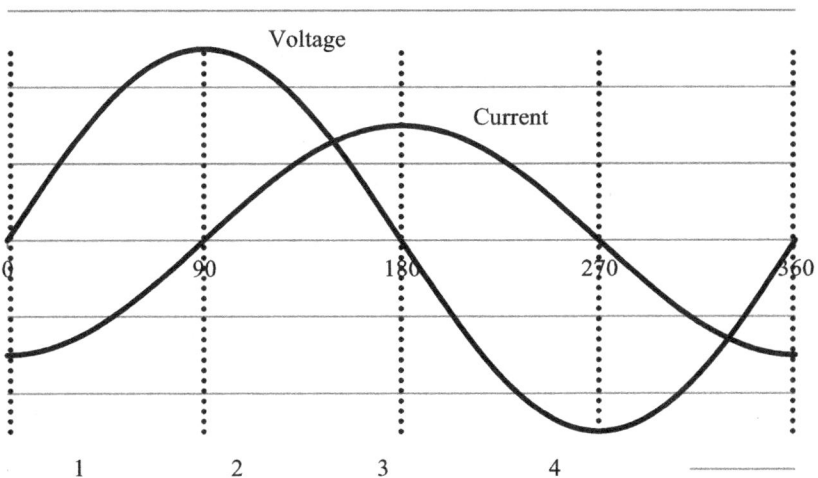

**Figure 7.4** A diagram showing the four combinations of voltage and current for an inductive load.

$$V_{av} = \frac{V_m}{\pi}(1 + \cos\alpha) \qquad (7.3)$$

From this equation, the contribution of each half-cycle to the overall output waveform can be calculated depending on the chosen delay angles. The frequency is adjusted in discrete amounts and is in proportion to the supply frequency.

As the single-phase cycloconverter has a positive converter and a negative converter, the operation of the cycloconverter is divided into four combinations (Figure 7.4 and Table 7.1).

**Table 7.1**

Four Combinations of Voltage and Current for the Operation of the Positive and Negative Converters with an Inductive Load

| Combination | Voltage | Current | Power | Positive Converter | Negative Converter |
|---|---|---|---|---|---|
| 1 | Positive | Negative | Negative | Inverting | Off |
| 2 | Positive | Positive | Positive | Rectifying | Off |
| 3 | Negative | Positive | Negative | Off | Inverting |
| 4 | Negative | Negative | Positive | Off | Rectifying |

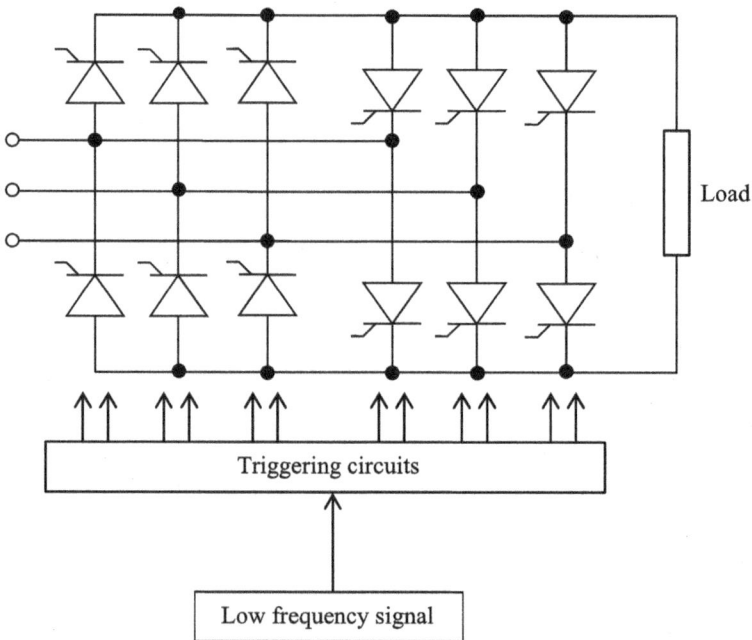

**Figure 7.5**  A three-phase to single phase cycloconverter.

## 7.2 Three-Phase Cycloconverter

A three-phase to single-phase cycloconverter consists of twelve thyristors arranged as two back-to-back converters (Figure 7.5). Three of these groups of twelve thyristors form a three-phase to three-phase cycloconverter (Figure 7.6). This number can be reduced to eighteen.

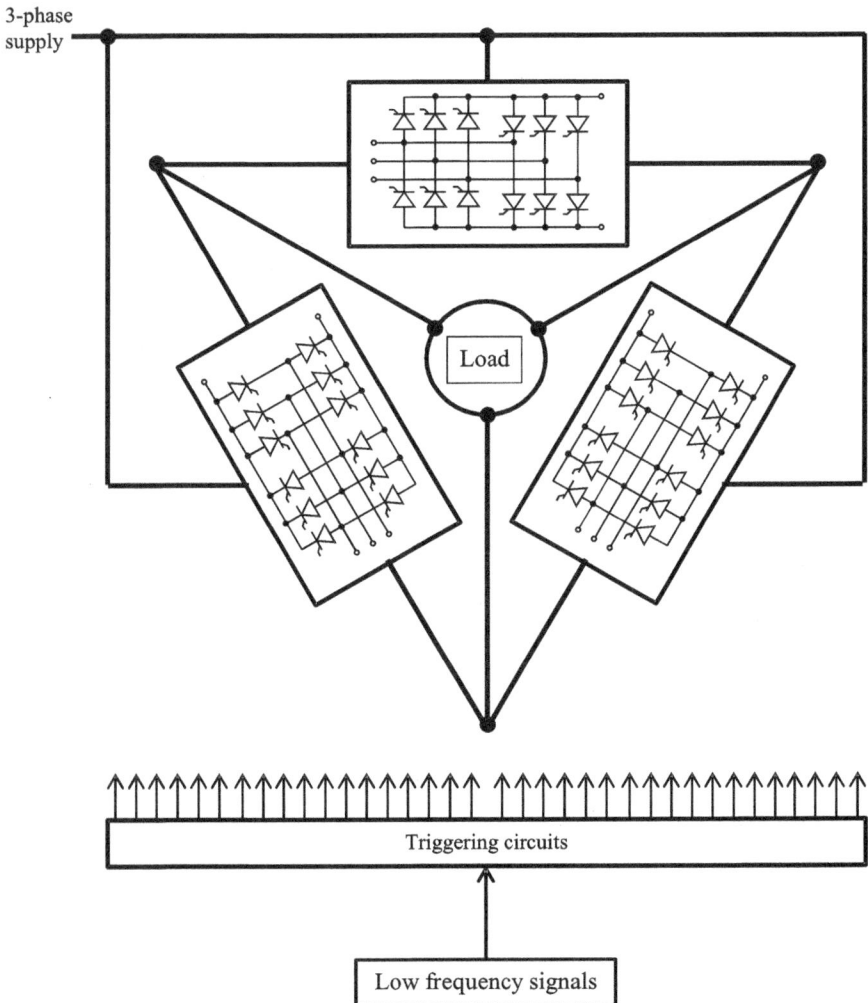

**Figure 7.6** Three-phase to three-phase cycloconverter with 36 thyristors.

The thyristors in a cycloconverter are naturally commutated and the circuit can feed either polarity of voltage with either polarity of current to the load. It has a four-quadrant capability when used in an electric drive application. The converter is a low-frequency (e.g., 60 Hz) to very low-frequency (e.g., less than 20 Hz) converter. The output voltage waveform becomes too distorted if the output frequency is set too high.

# 8

# Inverter

An inverter converts a dc voltage supply into an ac voltage. The source can be from a battery or from a rectified ac supply. If a thyristor converter is used to provide the dc supply voltage, then energy can be returned from the load to the ac supply. The design aim for an inverter is to synthesize a sinusoidal voltage waveform that has a controlled frequency and magnitude from a dc supply.

## 8.1 Single-Phase Inverter

In this converter, pairs of semiconductor devices are alternately turned on and off to generate a simple square wave with positive and negative polarity (Figures 8.1 and 8.2). Precautions must be taken to ensure that the power semiconductors in series across the dc supply are not conducting current at the same time. If $P_1$ is on, then it must be completely off before $N_2$ is turned on; otherwise, there will be a shoot-through of current with the possibility of the transistors being damaged. If the load has any inductive properties, then four diodes are needed to return the stored magnetic energy to the dc supply.

## 8.2 Three-Phase Inverter

Forming a three-phase inverter requires two extra devices (Figure 8.3). The top n-channel power semiconductors (labeled $P_1$, $P_2$, and $P_3$) can be replaced by p-channel devices should there be the availability of complementary pairs of n and p types. The phase voltages can be three simple square waves and are made by switching the devices in sequence (Figure 8.4).

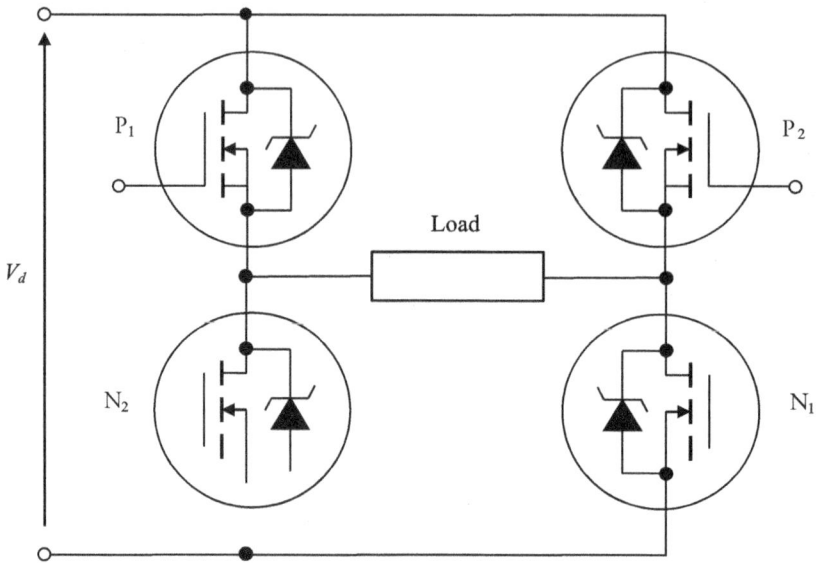

**Figure 8.1**   Single-phase inverter.

The line waveforms are six-step and an improvement from a simple square wave but are highly distorted. A better waveform is obtained by switching at a high frequency to synthesize a pulse width modulated sinusoidal waveform. The desired sinusoidal waveform (modulation) is compared to a higher-frequency triangular waveform (carrier) (Figure 8.5). Switching of the transistors occurs at each crossover of the two waveforms (Figure 8.6). When the magnitude of the triangular waveform is greater than that of the sinusoidal waveform, transistors $P_1$ and $N_1$ in Figure 8.1 are turned on and the supply voltage, $V_d$, is applied to

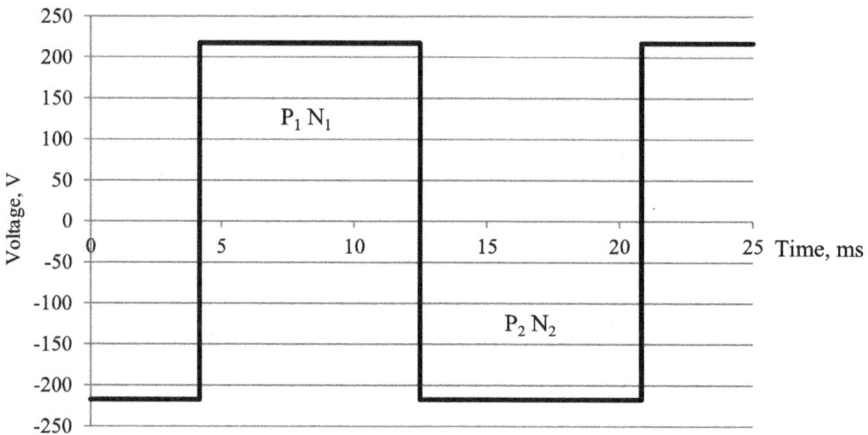

**Figure 8.2**   Inverter waveform for a 60-Hz and single-phase output voltage.

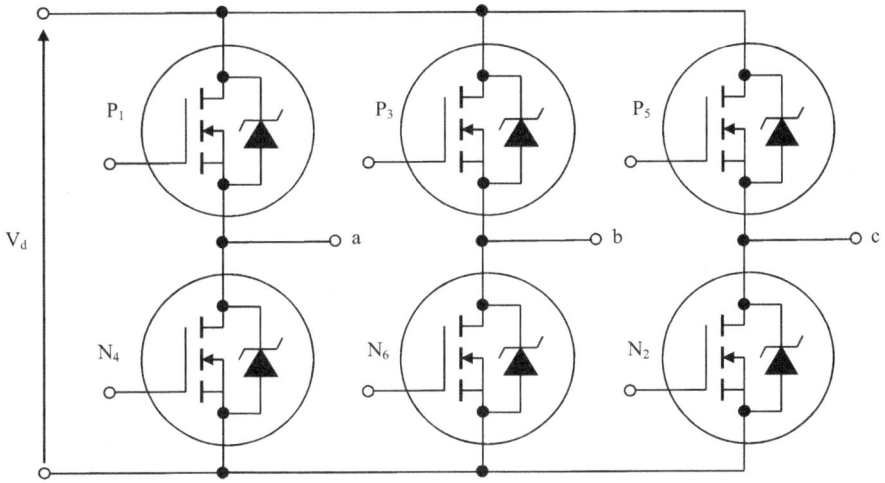

**Figure 8.3** Three-phase inverter.

the load. Transistors $P_1$ and $N_1$ are turned off and then $P_2$ and $N_2$ are turned on when the magnitude of the triangular waveform is less than that of the sinusoidal waveform. The output waveform operates at a constant frequency and alternates between $+V_d$ and $-V_d$ with a varying duty cycle When this pulse width modulation voltage waveform is applied to a typical inductive load the current

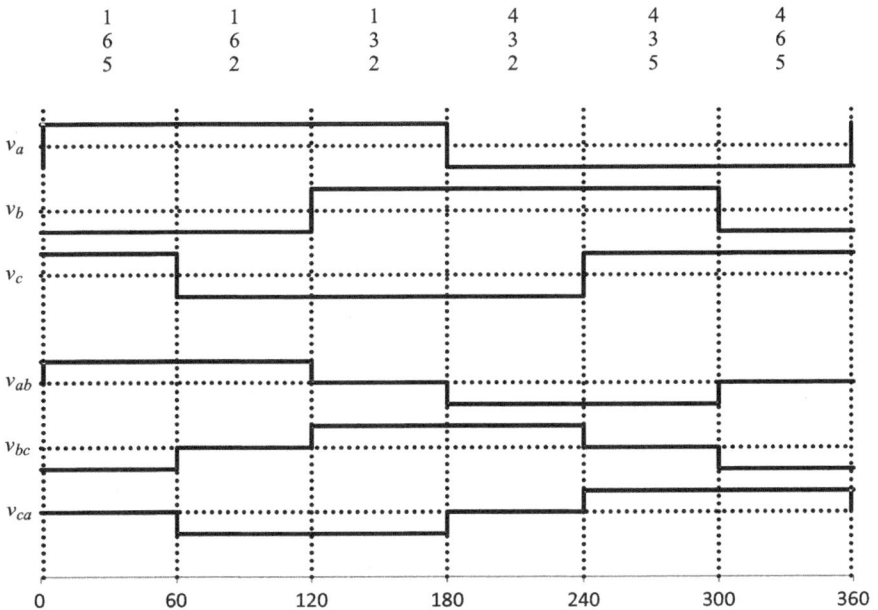

**Figure 8.4** Three-phase inverter waveforms.

has a sinusoidal appearance. As the applied voltage is constant and is either $+V_d$ or $-V_d$, then

$$\frac{di}{dt} = \frac{\pm V_d}{L} \qquad (8.1)$$

The current therefore rises and falls linearly with time at a constant rate (Figures 8.5 and 8.6). A current waveform with low distortion occurs when the triangular waveform has a high frequency compared to the desired sinusoidal waveform. Under these conditions, when the desired output current is at a low frequency, there are a large number of crossovers per cycle. The number of times that switching occurs during a cycle between $+V_d$ and $-V_d$ in the output pulse width modulation voltage waveform is high. With a constant triangular waveform frequency, as the desired output frequency increases, the number of transitions between $+V_d$ and $-V_d$ per cycle decreases with increased distortion of the current waveform. The maximum frequency of the triangular waveform and frequency of the pulse width modulation waveform have practical limits which limit the maximum frequency of the output current.

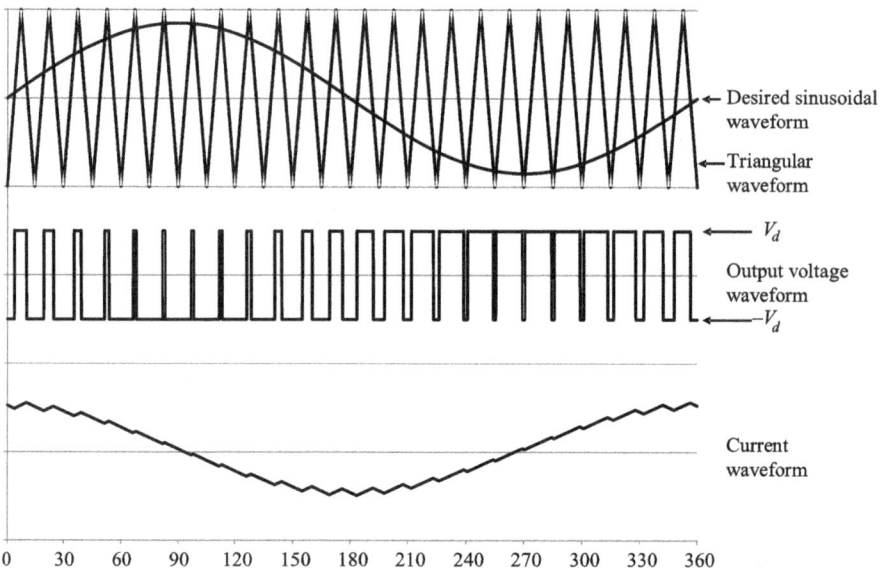

**Figure 8.5**   Single-phase pulse width modulation waveforms.

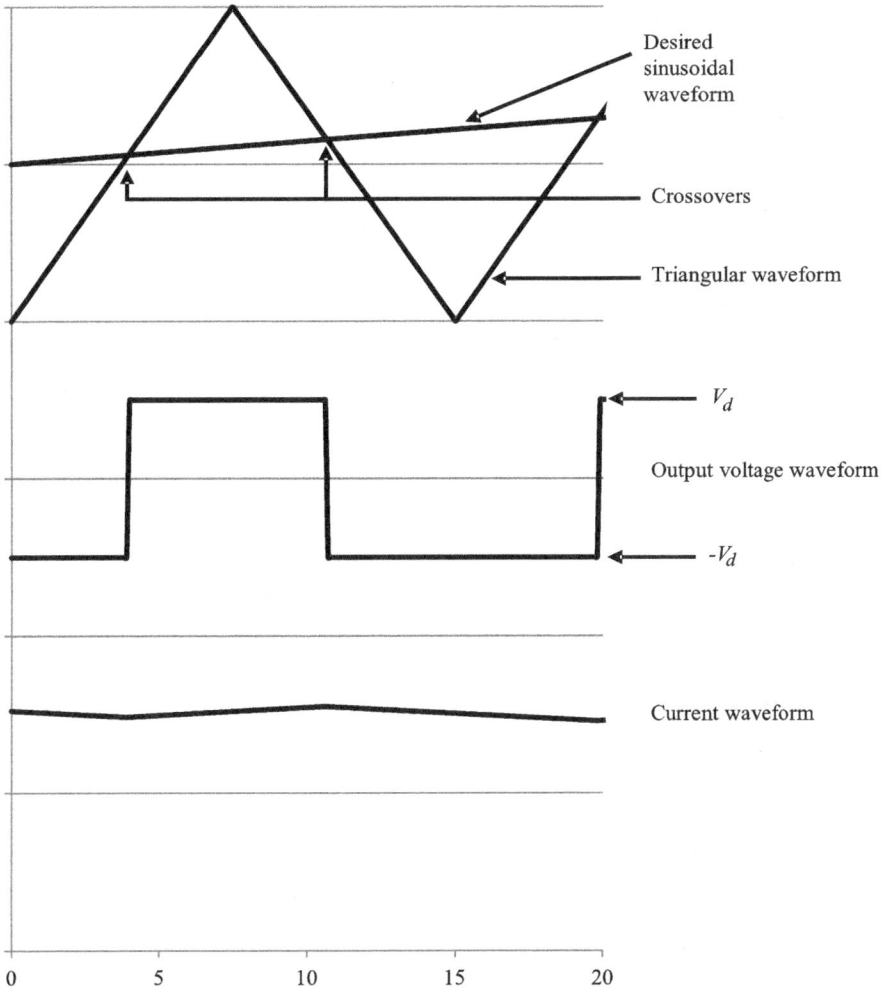

**Figure 8.6** Expanded view of single-phase waveforms shown in Figure 8.5.

Three modulation and carrier waveforms, which are phase shifted by 120°, are required for a three-phase inverter (Figure 8.7). For each intersection of the desired and modulated waveform with the triangular waveform, an analog comparator can be used to detect the crossovers and decode the switching sequence. A more practical implementation is to use a digital system with counters, timers, and lookup tables for the sinusoidal values. The pulse width modulation of sinusoidal voltages has application to the speed and torque control of induction motors.

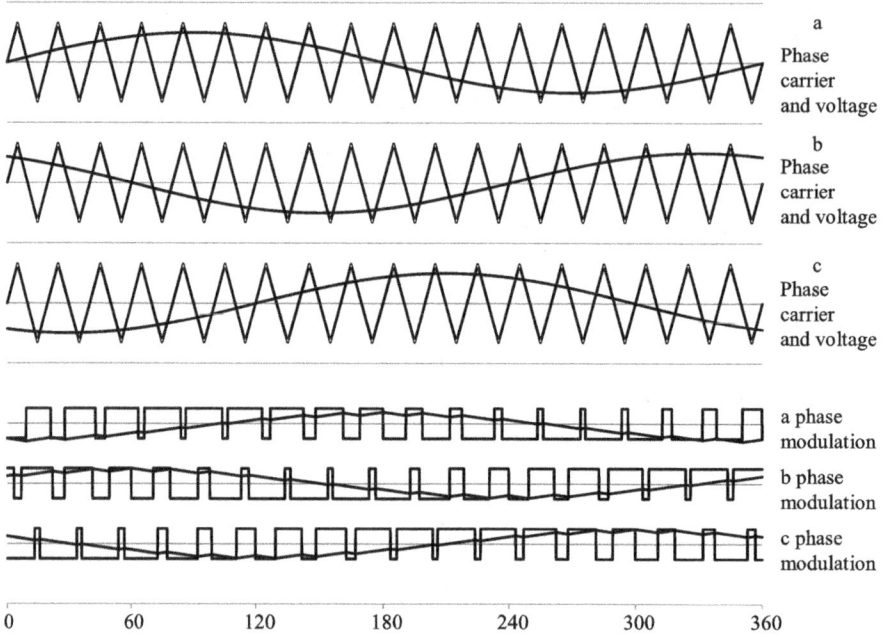

a
Phase
carrier
and voltage

b
Phase
carrier
and voltage

c
Phase
carrier
and voltage

a phase
modulation

b phase
modulation

c phase
modulation

0    60    120    180    240    300    360

**Figure 8.7**   Sinusoidal pulse width modulation waveforms with an inductive load.

# 9

# DC-to-DC Converters

## 9.1 Inductive Load

The average voltage across a load can be controlled by applying a dc supply voltage for some of the time at a fixed frequency. Applying or not applying a dc voltage is a process called chopping. A simple circuit for an inductive load can be made with a single transistor and freewheeling diode (Figure 9.1).

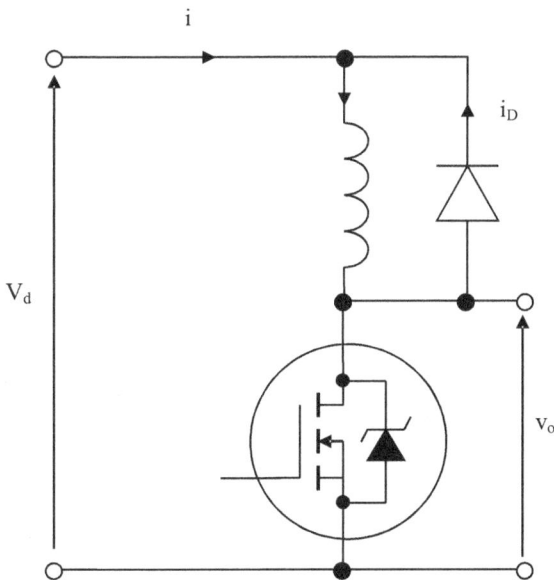

**Figure 9.1** Simple dc-to-dc converter with an inductive load and a free-wheeling diode.

Either polarity can be applied to a load with four switching devices (Figure 9.2).

The layout and interconnection of the components that form a power converter circuit have led historically to names being attributed to them. The circuit shown in Figure 9.2 is often called an H-bridge when used to control the output into a dc load. While the transistors are being switched and the output voltage is therefore ac, the output current is smoothed and dc. In this sense, the H-bridge is a dc-to-dc converter. In Figure 8.1, this circuit can convert a dc voltage to an ac voltage and then the circuit shown in Figure 9.2 is called an inverter circuit. The circuit converts dc power to ac power. It is not only the topology of a power electronic circuit that is important but also how it is operated and the application.

If a pair of transistors ($P_1$ and $N_1$ or $P_2$ and $N_2$) in Figure 9.2 are turned on and off at a fixed frequency into an inductive load, then the current waveform is triangular (Figure 9.3).

In Figure 9.3, the average voltage is given by

$$V_{av} = \frac{V_d\, t_1}{T_p} \tag{9.1}$$

or expressed in terms of the duty cycle

$$V_{av} = \delta V_d \tag{9.2}$$

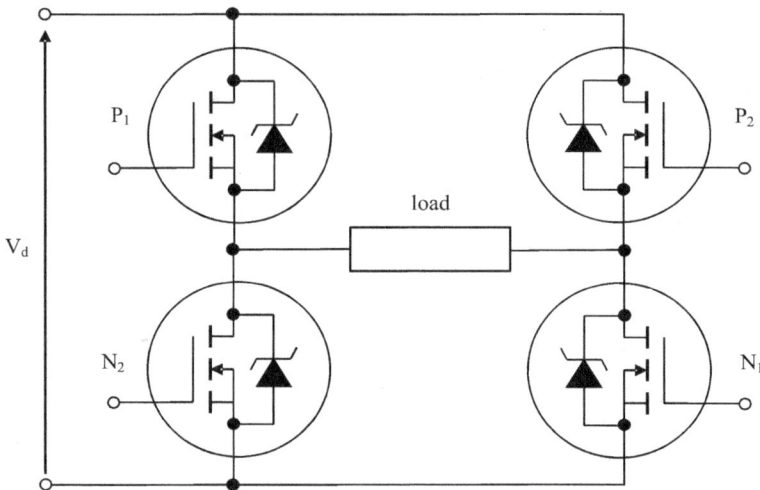

**Figure 9.2**  H-bridge. The top transistors can be either P-channel or N-channel types. Note that this circuit topology is the same as that for an inverter.

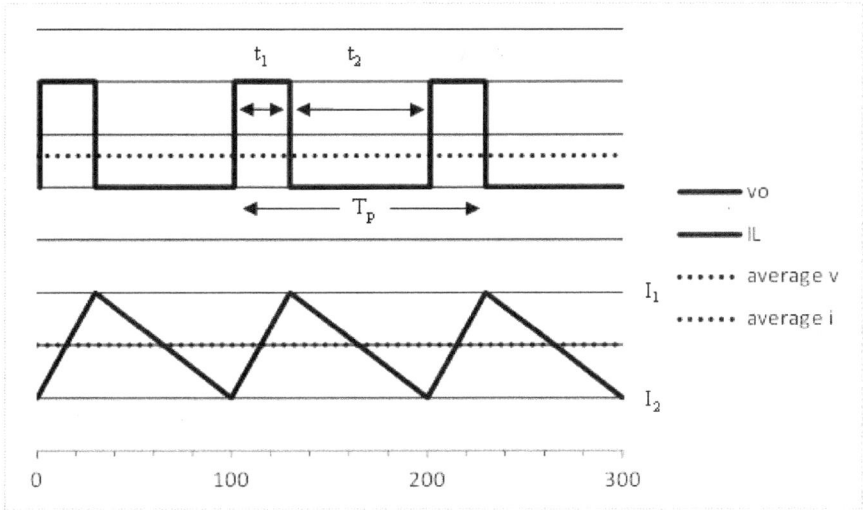

**Figure 9.3** Load voltage and current for a duty cycle of 0.3.

When $P_1$ and $N_1$ are turned on, the current in the inductor rises.

$$V_d = L\frac{di}{dt} \tag{9.3}$$

$$i = \int_0^t \frac{V_d}{L}\, dt \tag{9.4}$$

$$i = \frac{V_d}{L}t + I_k \tag{9.5}$$

where $I_k$ is the initial current.

The peak current, $I_1$, is given by

$$I_1 = \frac{V_d}{L}t_1 + I_k \tag{9.6}$$

When the transistors are turned off, the voltage across the load is $-V_d$ and the current falls. The current, $I_2$, at the end of the cycle is

$$I_2 = \frac{-V_d}{L}t_2 + I_k \tag{9.7}$$

Eliminating $I_k$ from (9.6) and (9.7),

$$I_1 = \frac{V_d}{L}t_1 + I_2 \tag{9.8}$$

If the current is steady, then the on and off times are equal.

$$I_2 = \frac{-V_d}{L} t_2 + I_1 \qquad (9.9)$$

$$I_1 = \frac{V_d}{L} t_1 + \frac{-V_d}{L} t_2 + I_1 \qquad (9.10)$$

$$t_1 = t_2 \qquad (9.11)$$

This analysis assumes that the active devices are ideal and do not have a voltage across them when they conduct.

The average current is simply

$$I_{av} = \frac{(I_1 + I_2)}{2} \qquad (9.12)$$

If the forward, $V_{d1}$, and reverse, $V_{d2}$, voltages are not the same, then for a steady current

$$\frac{V_{d1}}{-V_{d2}} = \frac{t_1}{t_2} \qquad (9.13)$$

An alternative circuit with two transistors requires a bipolar dc supply (Figure 9.4). This circuit can supply positive average voltages and negative voltages

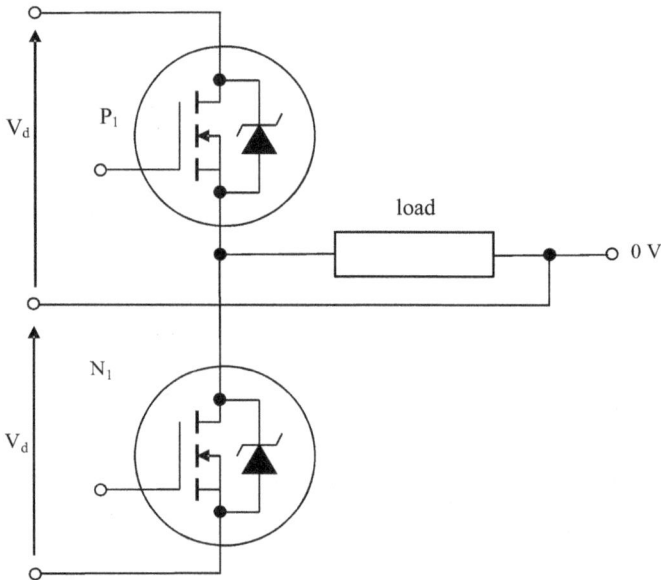

**Figure 9.4**    A dc-to-dc converter using a bipolar supply. The top transistor can be either a P-channel or N-channel device.

to the load. The duty cycle can be ranged from full positive voltage ($\delta = 1$) to full negative voltage ($\delta = -1$). The average voltage across the load is then

$$V_{av} = V_d (\delta - d_1 + d_2) \tag{9.14}$$

where $d_1$ and $d_2$ are the proportions of the cycle where there is zero voltage during the interlocking delays.

The waveform for the converter using a bipolar supply is shown in Figure 9.5. To avoid shoot-through where the two transistors are conducting at the same time, there are delays between the demand signals and when the transistors turn on or off. If the duty cycle is between 0 and 1, then the average voltage is positive, and when it is between 0 and −1, the average voltage is negative (Figure 9.6).

The forward and reverse dc supply voltages do not have to be equal. For a steady current the ratio forward and reverse times are given by (9.13) (assuming that there are no interlocking delays).

## 9.2 DC Machine

The control of a dc motor in the electric drives industry is wide ranging in powers and application. Electric vehicles require the control of torque and speed delivered from the motor to propel it forward and to return energy to the battery under deceleration. A simple circuit consists of one power semiconductor switch and a diode (Figure 9.7).

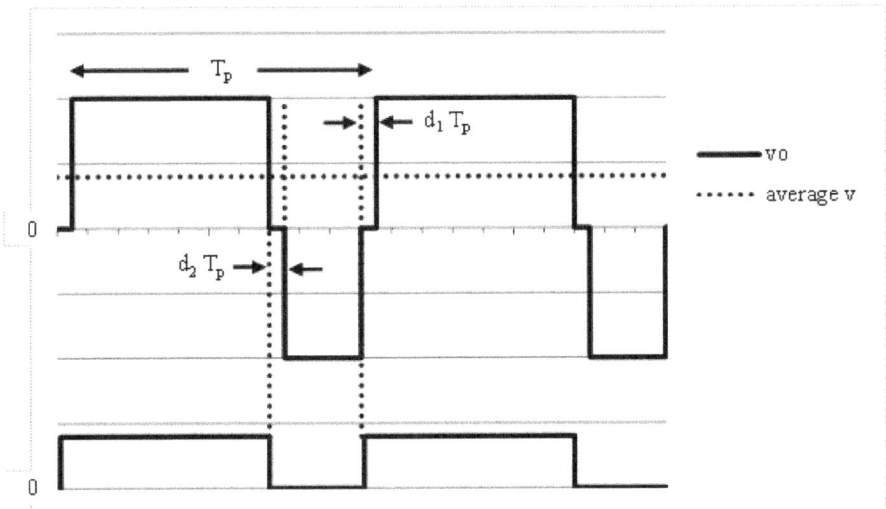

**Figure 9.5** Voltage waveform for the circuit in Figure 9.4, showing the effects of interlock delays. The average voltage is positive. The demand signal waveform is shown at the bottom.

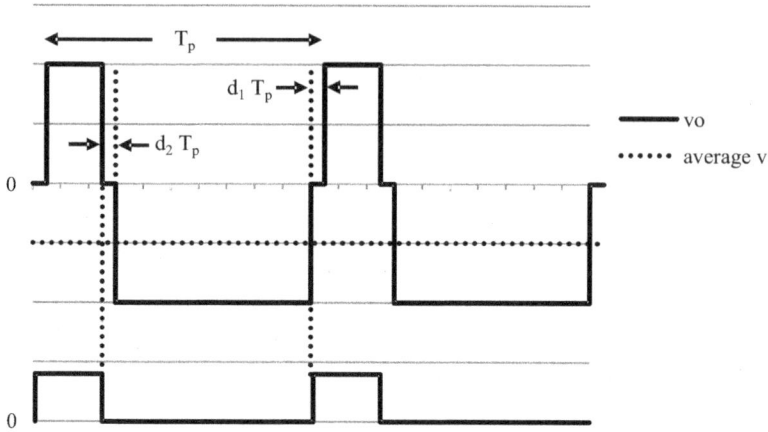

**Figure 9.6**  Voltage waveform for the circuit in Figure 9.4, showing the effects of interlock delays. The average voltage is negative. The demand signal waveform is shown at the bottom.

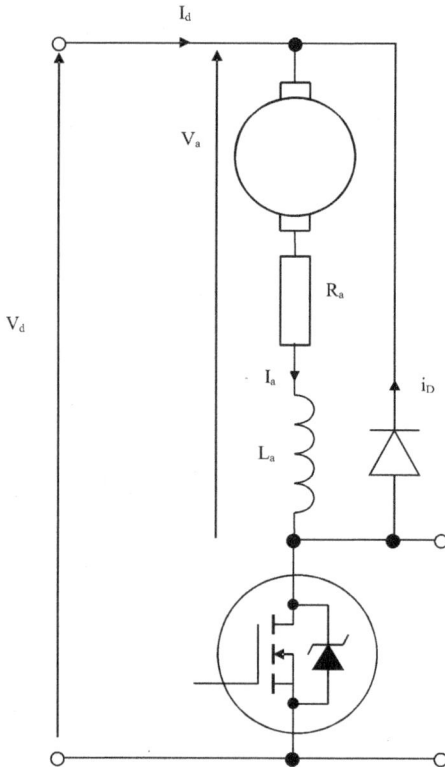

**Figure 9.7**  DC motor chopper circuit with freewheeling diode.

Ignoring any transients from the presence of the armature inductance, the dc operating conditions are as follows.

$$V_a = I_a R_a + k_m \omega \qquad (9.15)$$

$$V_a = \delta V_d \qquad (9.16)$$

$$\delta V_d = I_a R_a + k_m \omega \qquad (9.17)$$

where $I_a$, $R_a$, and $k_m$ are the armature current, armature resistance, and motor back emf constant, respectively. The mechanical output power is the product of the torque and rotational frequency so the efficiency, $\eta$, is

$$\eta = \frac{100 T_q \omega}{T_q \omega + I_a^2 R_a} \qquad (9.18)$$

The H-bridge circuit is very commonly used to control the positive and negative voltages to achieve forward and reverse rotation (Figure 9.8). In this application, the load consists of the armature inductance and resistance with the back emf voltage. There are several switching sequences that can be used (shown later in this chapter). In applications where the motor shaft position is controlled (servomotor) transistors, $P_1$ and $N_1$ are switched together and then $P_2$ and $N_2$ are switched together. The average voltage at the input terminals to the motor is zero for a 50% duty cycle. It is positive if the duty cycle is more than 50% and negative if it is less than 50%.

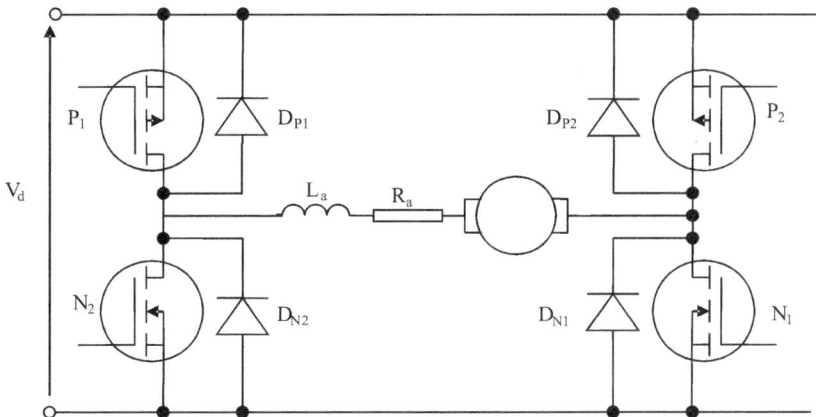

**Figure 9.8** H-bridge with a dc motor.

The H-bridge for the control of a dc motor has a simple circuit that is con-
structed with only four transistors and four diodes (Figure 9.8). However, how
it functions requires careful consideration to obtain the best performance from
the motor and circuit. In Figure 9.8, the transistors are shown separated from
their parasitic diodes. Discrete diodes are added if the main switching device is
a bipolar junction transistor. Turning on the transistors $P_1$ and $N_1$ allows current to flow from the power supply, $V_d$, to the motor. On turning off $P_1$ while
keeping $N_1$ on, the current flowing through the armature inductance, $L_a$, will
free-wheel through diodes $D_{N2}$ and $N_1$. Similarly, $N_1$ could be turned off and
current will then free-wheel through diodes $D_{N1}$ and $P_1$. If both $P_1$ and $N_1$ are
turned off together, then current will return to the supply via $D_{P2}$ and $D_{N2}$. In
this operation, positive voltage from the supply is applied to the motor allowing
it to rotate in a forward direction or develop positive torque.

Similarly, the motor can be rotated in the opposite direction by turning on
$P_2$ and $N_2$, in which case $D_{P1}$ and $D_{P2}$ act as the freewheeling or return diodes.
To avoid a shoot-through of current when a demand is made from the controller
to reverse the direction of rotation, some interlocking precautions are needed. If
$P_1$ is on and $N_2$ is turned on, then there will be a short circuit across the supply
resulting in either or both transistors being destroyed. To avoid this situation,
a predetermined time delay is applied between turning off $P_1$ and turning on
$N_2$. The current in $P_1$ should ideally cease before $N_2$ is turned on, requiring the
monitoring of the current through the transistors.

An alternative operation is to alternate the switching of the pairs of transistors $P_1$ with $N_1$ and $P_2$ with $N_2$. In this method, positive and negative voltages are applied across the motor terminals. Shoot-through must be avoided by
applying a dead band between the switching of the pairs of transistors.

Controlling the dc motor successfully requires information to be sent from
an electronic controller to the gates of the transistors in the H-bridge. Table 9.1
shows the 16 possible logical combinations of the states for the transistors.

During the rotation of the motor shaft or when developing torque, there
will be current flowing in the motor inductance and hence stored energy. Under
PWM control, when the circuit switches into either of the states $s_0$, $s_1$, $s_2$, $s_4$, or $s_8$,
there is stored magnetic energy in the inductor that will be returned to the supply
terminals and dissipated in the transistors, diodes, or motor resistance. Switching
between states $s_{12}$ and $s_4$ allows for a positive voltage to be applied to the motor
(Figures 9.9 and 9.10). In state $s_4$, the motor current freewheels through $N_1$ and
$D_{N2}$. An alternative operation is to switch between states $s_{12}$ and $s_8$, which allows
for current to freewheel through $P_1$ and $D_{P2}$ (Figure 9.11). In state $s_0$ current
will flow through $D_{P2}$ and $D_{N2}$, returning energy to the supply (Figure 9.12). In
logic state $s_5$, where $N_1$ and $N_2$ are on together, there is a short circuit across the
motor terminals (Figure 9.13).

**Table 9.1**

Logic States for the Transistors Shown in Figure 9.10

| State | Comment | $P_1$ | $N_1$ | $P_2$ | $N_2$ | $D_{P1}$ | $D_{N1}$ | $D_{P2}$ | $D_{N2}$ | Corresponding Figure |
|---|---|---|---|---|---|---|---|---|---|---|
| $s_0$ | No supply to the motor | 0 | 0 | 0 | 0 | $f_b$ | $f_b$ | $r_b$ | $r_b$ | Figure 9.17 |
| | | | | | | $r_b$ | $r_b$ | $f_b$ | $f_b$ | Figure 9.12 |
| $s_1$ | No supply freewheel | 0 | 0 | 0 | 1 | $r_b$ | $r_b$ | $f_b$ | $f_b$ | Figure 9.15 |
| $s_2$ | No supply freewheel | 0 | 0 | 1 | 0 | $r_b$ | $r_b$ | $f_b$ | $f_b$ | Figure 9.16 |
| $s_3$ | Reverse motor | 0 | 0 | 1 | 1 | $r_b$ | $r_b$ | $r_b$ | $r_b$ | Figure 9.14 |
| $s_4$ | No supply freewheel | 0 | 1 | 0 | 0 | $f_b$ | $f_b$ | $r_b$ | $r_b$ | Figure 9.10 |
| $s_5$ | Short-circuit motor | 0 | 1 | 0 | 1 | $r_b$ | $r_b$ | $r_b$ | $f_b$ | Figure 9.13 |
| | | | | | | $r_b$ | $f_b$ | $r_b$ | $r_b$ | |
| $s_6$ | Shoot-through | 0 | 1 | 1 | 0 | | | | | |
| $s_7$ | Shoot-through | 0 | 1 | 1 | 1 | | | | | |
| $s_8$ | No supply freewheel | 1 | 0 | 0 | 0 | $f_b$ | $f_b$ | $r_b$ | $r_b$ | Figure 9.11 |
| $s_9$ | Shoot-through | 1 | 0 | 0 | 1 | | | | | |
| $s_{10}$ | Short-circuit motor | 1 | 0 | 1 | 0 | $r_b$ | $r_b$ | $f_b$ | $r_b$ | |
| | | | | | | $f_b$ | $r_b$ | $r_b$ | $r_b$ | Figure 9.18 |
| $s_{11}$ | Shoot-through | 1 | 0 | 1 | 1 | | | | | |
| $s_{12}$ | Forward motor | 1 | 1 | 0 | 1 | $r_b$ | $r_b$ | $r_b$ | $r_b$ | Figure 9.19 |
| $s_{13}$ | Shoot-through | 1 | 1 | 0 | 1 | | | | | |
| $s_{14}$ | Shoot-through | 1 | 1 | 1 | 0 | | | | | |
| $s_{15}$ | Shoot-through | 1 | 1 | 1 | 1 | | | | | |

*Note:* 0 indicates that a transistor is on and 1 indicates that it is off. The states of the diodes are shown in the right columns as either forward-biased ($f_b$) or reverse-biased ($r_b$).

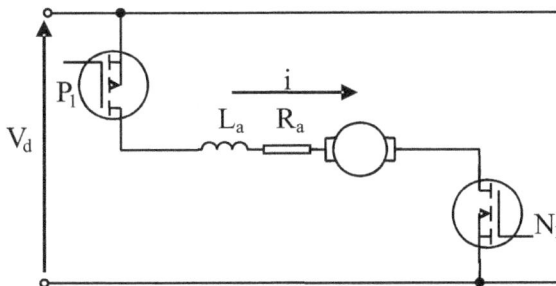

**Figure 9.9**  Motoring logic state $s_{12}$.

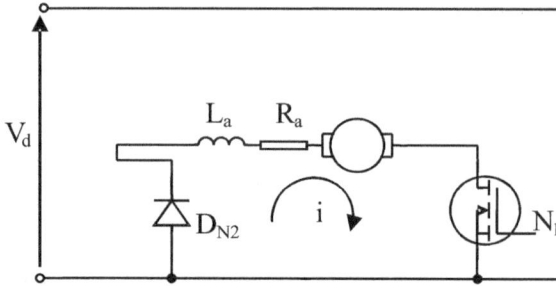

**Figure 9.10**   Freewheeling current logic state $s_4$.

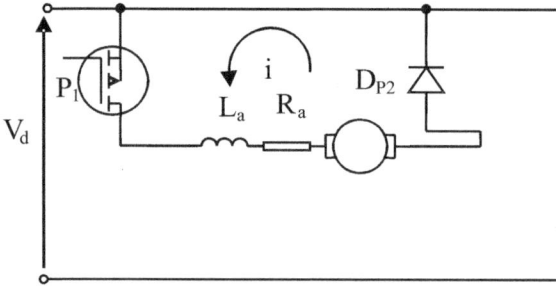

**Figure 9.11**   Freewheeling current logic state $s_8$.

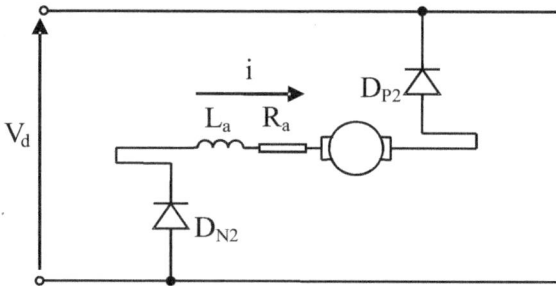

**Figure 9.12**   Current returning to the supply logic state $s_0$.

In the reverse direction, similar transistors and diodes combinations are conducting current. Table 9.1 shows the active components. In either forward or reverse operation when the two transistors are on in logic state $s_5$ or $s_{10}$ (Figures 9.13 and 9.18), current can flow in either direction through the motor and power is dissipated in the motor resistance and active transistors.

If the motor is running at full speed under no load conditions, then the back emf may be at its maximum. If a demand is made that requires instantaneous reversal of the motor, then the H-bridge will be switched from state $s_{12}$ to $s_3$ or vice versa. The voltage in the circuit is then the series combination of the

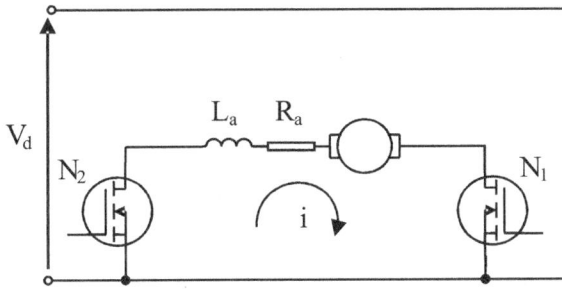

**Figure 9.13**  Short-circuit current logic state $s_5$.

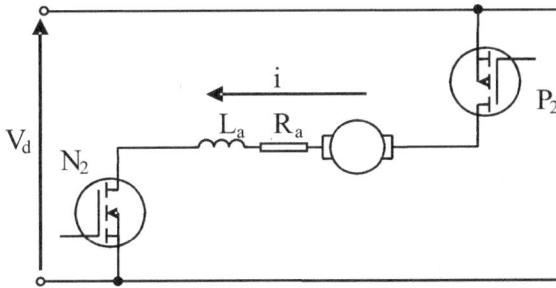

**Figure 9.14**  Reverse motoring logic state $s_3$.

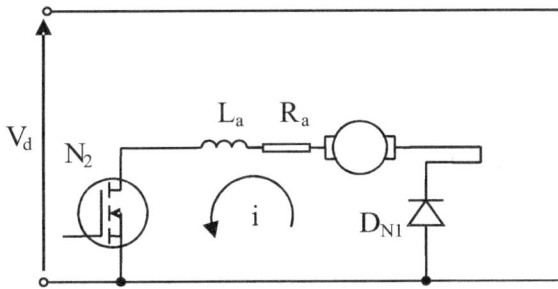

**Figure 9.15**  Freewheeling current logic state $s_1$.

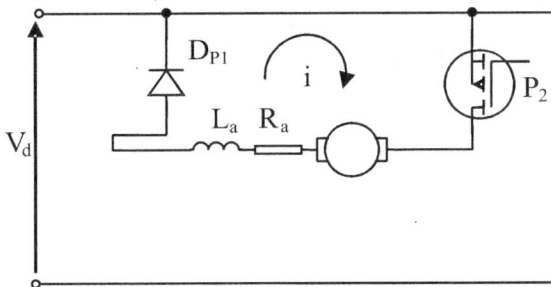

**Figure 9.16**  Freewheeling current logic state $s_2$.

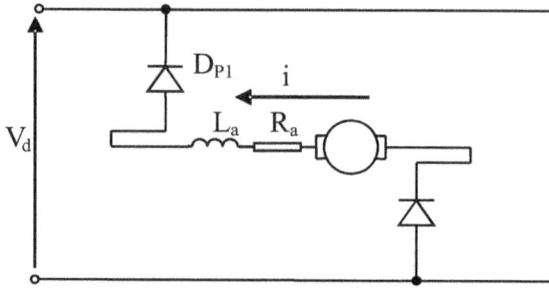

**Figure 9.17**   Current returning to the supply logic state $s_0$.

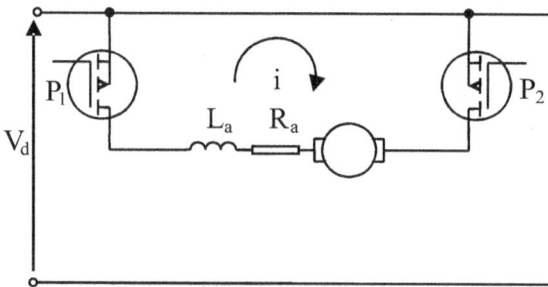

**Figure 9.18**   Short-circuit current logic state $s_{10}$.

dc supply voltage and the back emf. As the back emf will be close in value to the supply voltage, there is in effect twice the supply voltage in the circuit. This may lead to excessive currents in the semiconductor devices, in particular, in the diodes and possible damage to the motor.

## 9.3   Regeneration

Using a transistor and diode, energy stored in a mechanical load can be transferred to the electrical supply (Figure 9.19). Turning on the transistor reverse-biases the diode and places a short circuit across the armature of the dc motor (Figure 9.19). The armature current increases to a point when the transistor is turned off. The voltage across the transistor rises until the diode is forward-biased and energy from the inductance is returned to the supply. This circuit can be used during braking when the kinetic energy in a vehicle can be recovered and stored in a battery (kinetic energy recovery system, KERS). A simpler method is to use a resistor and a transistor to dissipate the mechanical energy as heat (Figure 9.20).

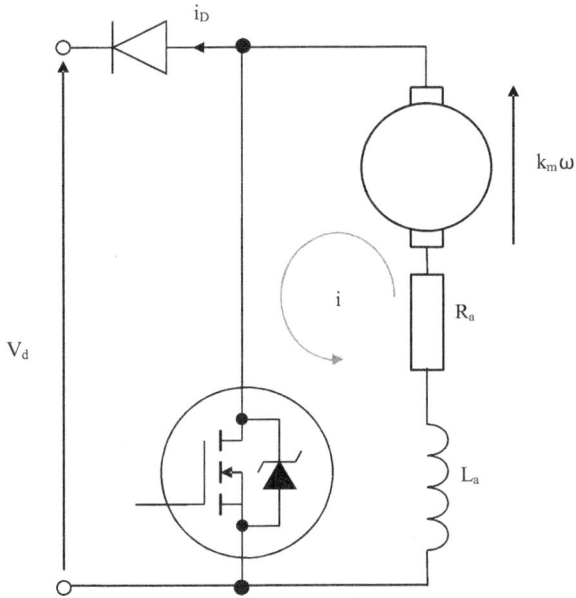

**Figure 9.19** Regeneration of mechanical energy returning to the electrical supply.

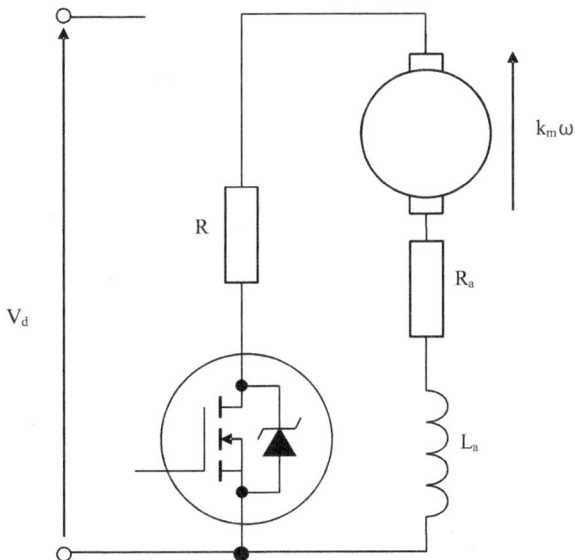

**Figure 9.20** Dissipation of mechanical energy in the resistor, *R*.

## 9.4   Step-Up and Step-Down DC-to-DC Converters

There are several types of dc-to-dc converters that directly transfer energy from either a low voltage to a high voltage or vice versa. The main types are step-up or boost, step-down, or buck and Ćuk.

The basic principle of operation of these circuits is to turn on a transistor, thereby storing energy in an inductor and then to turn off the transistor, which transfers the stored magnetic energy via a diode to the output load. The switching frequency is much higher than that of ac power supplies (60 or 50 Hz) and is typically in the tens of kilohertz. Operation at a high frequency offers the advantage of using a small inductor. There are two modes of operation where the current is either continuous or discontinuous. It is useful to note the conditions at the boundary between these two modes. Table 9.2 shows the assumptions to formulate equations describing the operation of the circuits.

With a large load, the current is continuous and above zero all the time (Figure 9.21). As the load decreases, the current falls to a situation where just before the transistor is turned on the current just touches zero and immediately starts to rise (Figure 9.22). The current is then at the boundary between continuous and discontinuous operation. At small loads the current can stay at zero for part of the period and is in discontinuous operation (Figure 9.23).

## 9.5   Step-Down (Buck) Converter

The basic design of the step-down converter consists of a switching device, a diode, and an inductor (Figure 9.24). When the transistor is turned on, the current in the inductor and load rise as the voltage across the inductor is the difference between the input and output voltages, $(V_d - V_o)$ (upper circuit in Figure 9.25). On turning off the transistor, the current falls as the voltage across the inductor is the reverse polarity of the output voltage, $-V_o$. The diode is forward-biased and maintaining the voltage at one terminal of the inductor to the voltage at the negative terminal of the dc supply (lower circuit in Figure 9.25).

**Table 9.2**
Assumptions and Approximations Used in an Analysis of DC-to-DC Converters

| |
|---|
| 1. The input voltage is constant. |
| 2. There is zero voltage across the transistor when it is conducting. |
| 3. The transistor has zero switching losses. |
| 4. The output voltage is constant (large capacitor). |
| 5. There is zero voltage across the diode when it is forward-biased. |

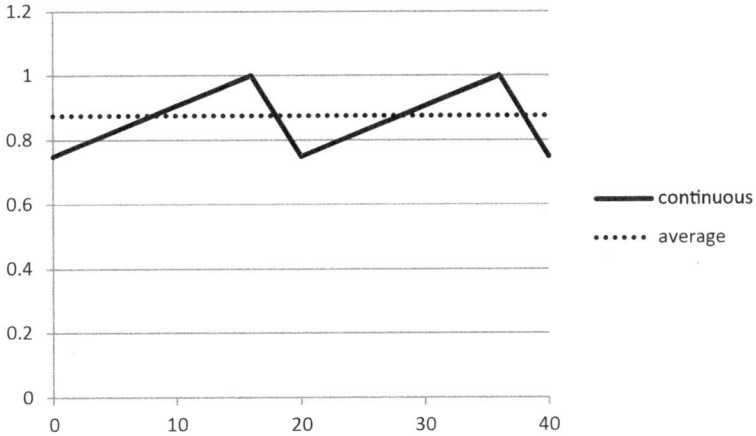

**Figure 9.21**  Continuous current operation.

As the current is still flowing in the direction established when the transistor was switched on, then the output voltage is negative with respect to the negative terminal of the dc supply. The magnitude of the output voltage is lower than that of the input voltage. The circuit has two states as shown in Figure 9.25.

The average voltage across the inductor is zero (lower waveform in Figure 9.26) for steady state operation and continuous current, otherwise the average current will rise or fall (upper waveform in Figure 9.26).

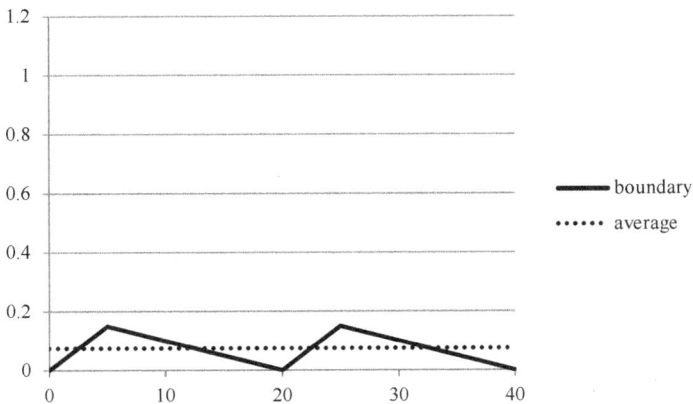

**Figure 9.22**  Current at the boundary between continuous and discontinuous operations.

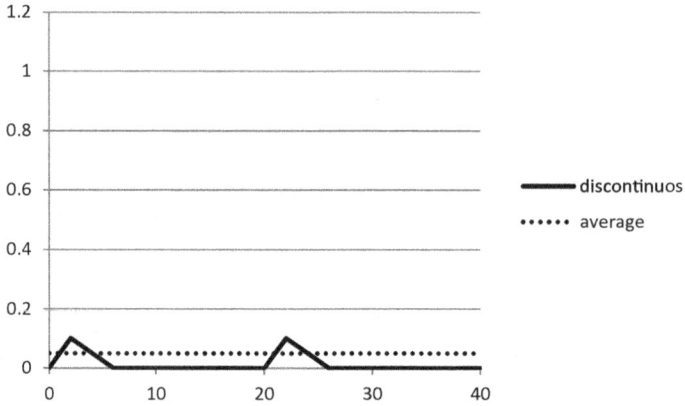

**Figure 9.23**  Discontinuous current operation.

The average voltage, $v_L$, over the period, $T_p$, is given by

$$V_{av} = \int_0^{T_p} v_L \, dt \tag{9.19}$$

$$\int_0^{T_p} v_L \, dt = \int_0^{t_1} v_L \, dt + \int_{t_1}^{t_2} v_L \, dt \tag{9.20}$$

As the average voltage is zero

$$\int_0^{t_1} v_L \, dt = -\int_{t_1}^{t_2} v_L \, dt \tag{9.21}$$

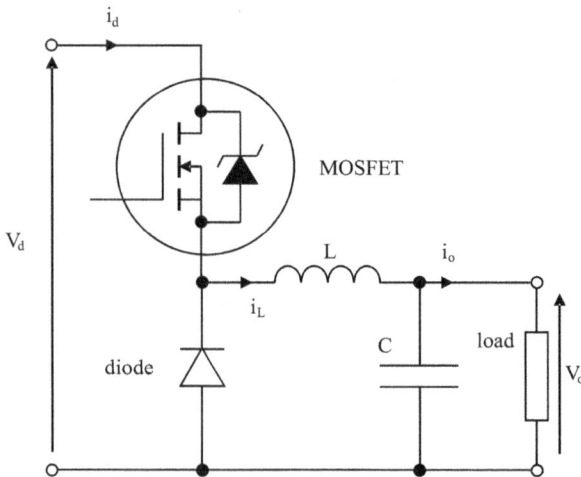

**Figure 9.24**  Step-down (buck) dc-to-dc converter.

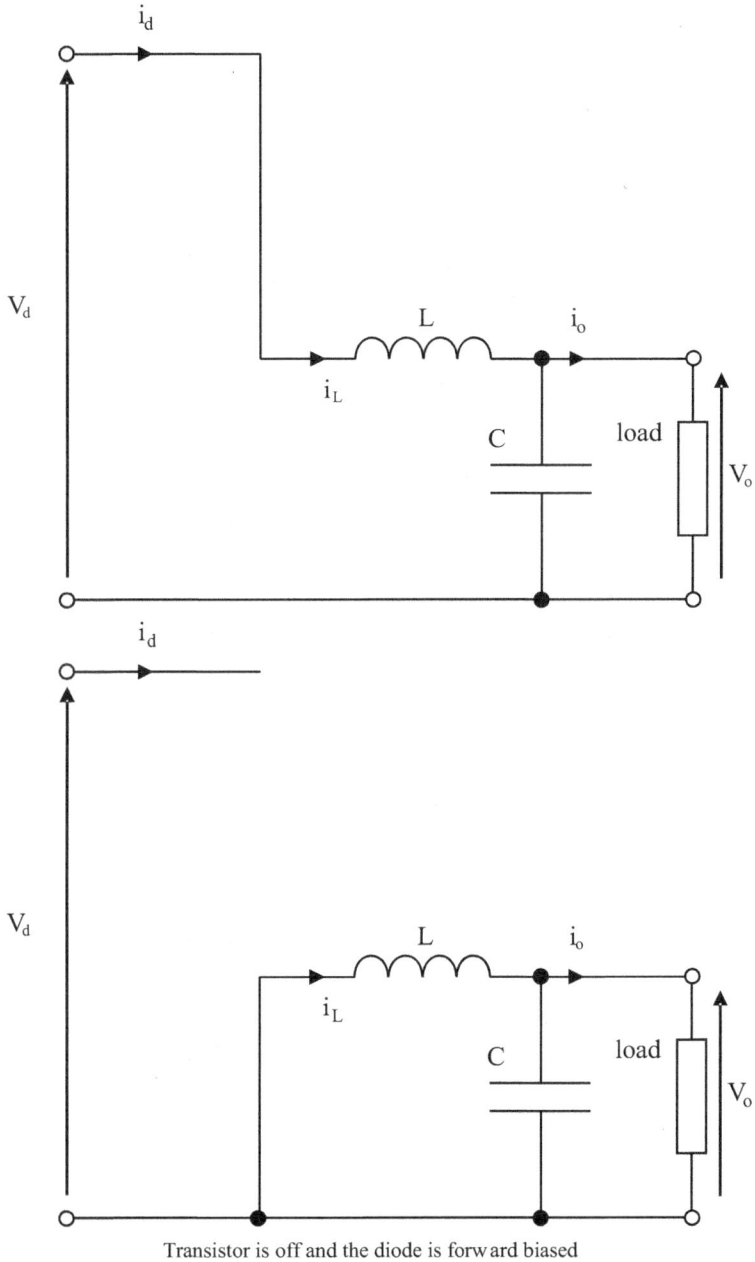

Transistor is off and the diode is forward biased

**Figure 9.25** The two states of the step-down circuit with an ideal transistor and diode.

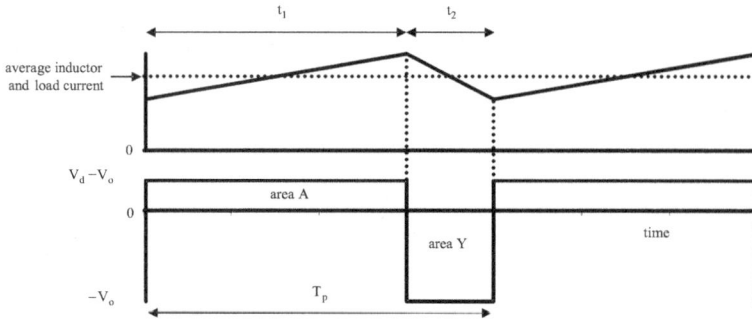

**Figure 9.26**   Current waveforms for a step-down dc-to-dc converter. The triangular inductor and load currents at are shown at the top. The inductor voltage is shown at the bottom.

The integrals are represented in the voltage waveforms as voltage-time areas (areas $X$ and $Y$).

$$(V_d - V_o)t_1 = V_o t_2 \tag{9.22}$$

This relationship between the input and output voltages can be expressed in terms of the duty cycle.

$$V_d t_1 = V_o (t_1 + t_2) \tag{9.23}$$

$$\delta = \frac{t_1}{T_P} = \frac{V_o}{V_d} \tag{9.24}$$

Therefore, the duty cycle sets the ratio between the input and output voltages. The aim of a dc-to-dc step-down converter is to maintain a constant output voltage, $V_o$, regardless of any variation in the input voltage, $V_d$. The output voltage is maintained constant by adjustments in the duty cycle.

For a triangular waveform the average current is

$$I_{av} = \frac{I_1 + I_2}{2} \tag{9.25}$$

At the boundary between continuous and discontinuous current the current at the start and end of the cycle is zero ($I_1 = 0$). The average inductor current at the boundary, $I_{Lb}$, is

$$I_{Lb} = \frac{I_2}{2} \tag{9.26}$$

During the time that the transistor is on, $t_1$

$$(V_d - V_o) = L\frac{I_2}{t_1} \tag{9.27}$$

Therefore,

$$I_{Lb} = \frac{t_1}{2L} + (V_d - V_o) \qquad (9.28)$$

or

$$I_{Lb} = \frac{\delta\, T_p}{2L} (V_d - V_o) \qquad (9.29)$$

Because the inductor current and output current are equal, then the average output current at the boundary, $I_{ob}$, is

$$I_{ob} = \frac{\delta\, T_p}{2L} (V_d - V_o) \qquad (9.30)$$

$$I_{ob} = \frac{\delta\, T_p}{2L} (V_d - \delta V_d) \qquad (9.31)$$

$$I_{ob} = \frac{\delta\, T_p V_d}{2L} (1 - \delta) \qquad (9.32)$$

An example of the variation in output voltage (9.24) and current (9.31) is plotted in Figure 9.27 for the values of the duty cycle.

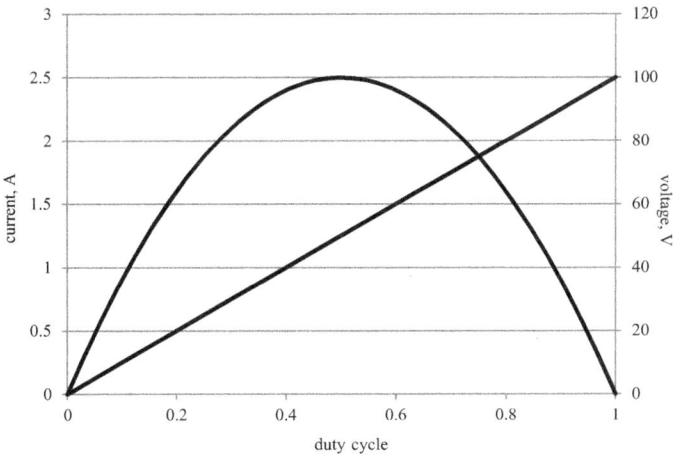

**Figure 9.27**  Inductor or output current curve (at the boundary of continuous and discontinuous current) and linear output voltage as a function of duty cycle for a 50-kHz and 100-V input dc-to-dc step-down converter with a 100-μH inductor.

This characteristic for the average currents reaches a maximum at a duty cycle of 0.5 when the current is given by

$$I_{ob} = \frac{0.125\, T_p V_d}{L} \tag{9.33}$$

A plot of the input current and voltage is shown in Figure 9.28 while keeping the output voltage constant. Under discontinuous conditions, the current falls to zero in a time of $\gamma T_p$ (Figure 9.29). It is then zero for a time $(1 - \delta - \gamma)T_p$.

As the current is steady, then

$$(V_d - V_o)\delta T_p = V_o \gamma T_p \tag{9.34}$$

$$\frac{V_o}{V_d} = \frac{\delta}{(\gamma + \delta)} \tag{9.35}$$

The peak inductor current is

$$I_2 = \frac{(V_d - V_o)\delta T_p}{L} \tag{9.36}$$

$$I_2 = \frac{V_o \gamma T_p}{L} \tag{9.37}$$

$$\frac{V_o \gamma T_p}{L} = \frac{(V_d - V_o)\delta T_p}{L} \tag{9.38}$$

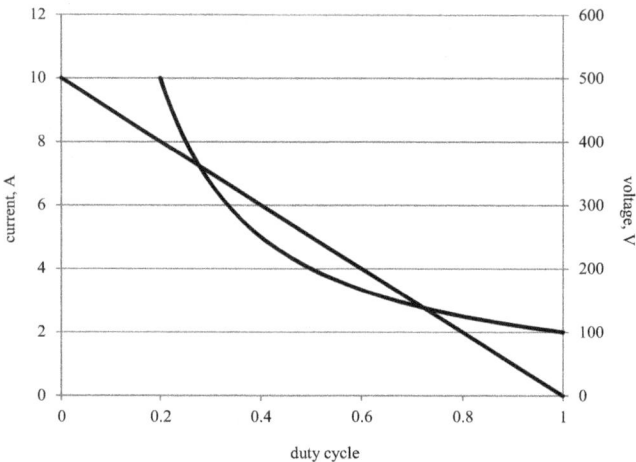

**Figure 9.28**   Inductor or output current curve and linear input voltage as a function of duty cycle for a 50-kHz and 100-V input dc-to-dc step-down converter with a 100-µH inductor.

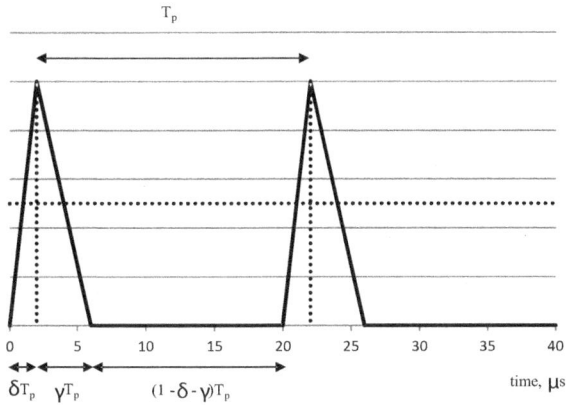

**Figure 9.29**   Discontinuous current in a step-down converter.

$$\gamma = \delta \frac{V_d - V_o}{V_o} \tag{9.39}$$

The average inductor current, $I_L$, is

$$I_L = \frac{I_2(\delta T_p) + I_2(\gamma T_p)}{2T_p} \tag{9.40}$$

$$I_L = \frac{I_2(\gamma + \delta)}{2} \tag{9.41}$$

$$I_L = \frac{(V_d - V_o)\delta T_p \ (\gamma + \delta)}{2L} \tag{9.42}$$

$$\gamma + \delta = \delta \frac{V_d - V_o}{V_o} + \delta \tag{9.43}$$

$$\gamma + \delta = \delta \frac{V_d}{V_o} \tag{9.44}$$

$$I_L = \frac{(V_d - V_o)\delta T_p}{2L} \delta \frac{V_d}{V_o} \tag{9.45}$$

$$I_L \ \frac{\delta^2 T_p (V_d - V_o)V_d}{2LV_o} \tag{9.46}$$

The curves for continuous and discontinuous current are plotted in Figure 9.30 for ratios of input to output voltages using (9.24) and (9.45). The boundary between these two operating regimes is plotted from (9.31). Note the horizontal lines in continuous current operation because the duty cycle sets voltage ratio and is independent of the current.

## 9.6  Step-Up (Boost) Converter

Turning on the transistor in Figure 9.31 applies the input voltage across the inductor and the diode is reverse-biased (upper circuit in Figure 9.32). Current flows from the inductor through the diode to the output load. The diode becomes forward-biased on turning off the transistor and energy stored in the inductor is transferred to the output (lower circuit in Figure 9.32). The output voltage is then higher than the input voltage.

Under continuous current operation and at sufficiently high frequency, the current waveform is triangular (Figure 9.33).

The average voltage, $V_{av}$, over the period, $T_p$, is given by

$$V_{av} = \int_0^{T_p} v_L dt \tag{9.47}$$

$$\int_0^{T_p} v_L dt = \int_0^{t_1} v_L dt + \int_{t_1}^{t_2} v_L dt \tag{9.48}$$

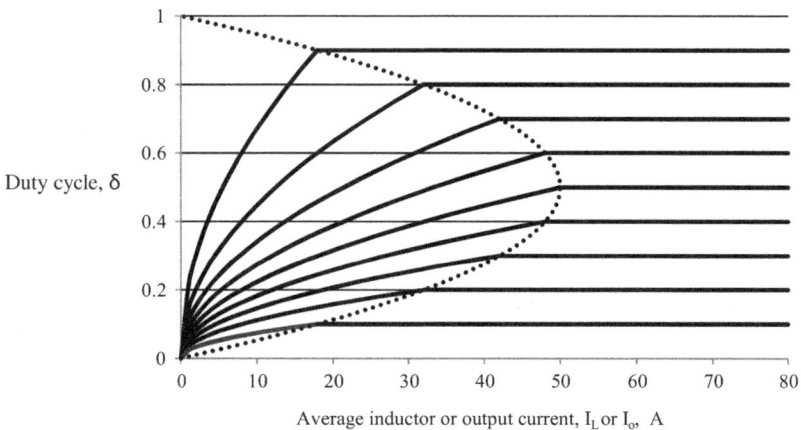

**Figure 9.30**  Continuous current on the right and discontinuous current on the left with the boundary (dotted curve). The ratios of input to output voltages ($V_o/V_d$) range from 0.9 to 0.1 in steps of 0.1 (top to bottom) ($L = 5 \, \mu H$, $f = 50$ kHz, $V_d = 100$V).

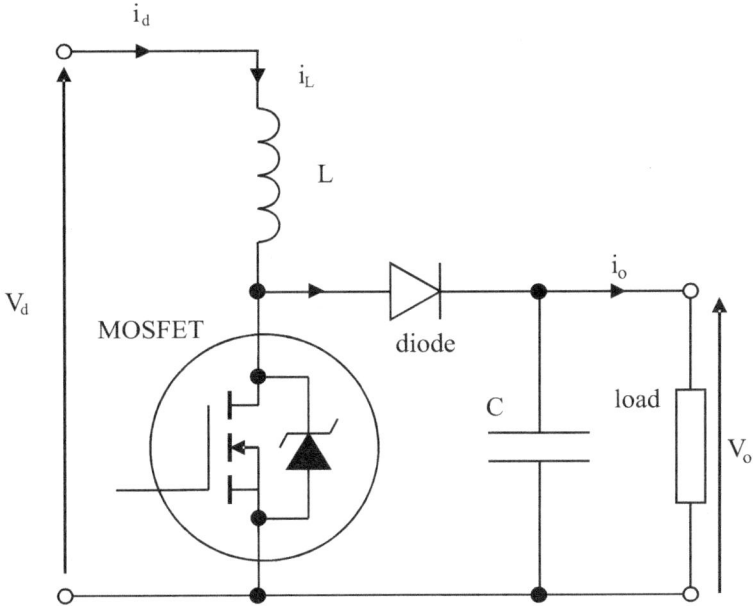

**Figure 9.31** Step-up (boost) dc-to-dc converter.

As the average voltage is zero

$$\int_0^{t_1} v_L dt = -\int_{t_1}^{t_2} v_L dt \qquad (9.49)$$

The integrals are represented in the voltage waveforms as voltage-time areas (areas $X$ and $Y$).

$$V_d t_1 = -(V_d - V_o) t_2 \qquad (9.50)$$

This relationship between the input and output voltages can be expressed in terms of the duty cycle.

$$V_d (t_1 + t_2) = V_o t_2 \qquad (9.51)$$

$$\frac{V_o}{V_d} = \frac{T_p}{t_2} = \frac{1}{1-\delta} \qquad (9.52)$$

$$\delta = \frac{t_1}{T_p} = \frac{V_o - V_d}{V_o} \qquad (9.53)$$

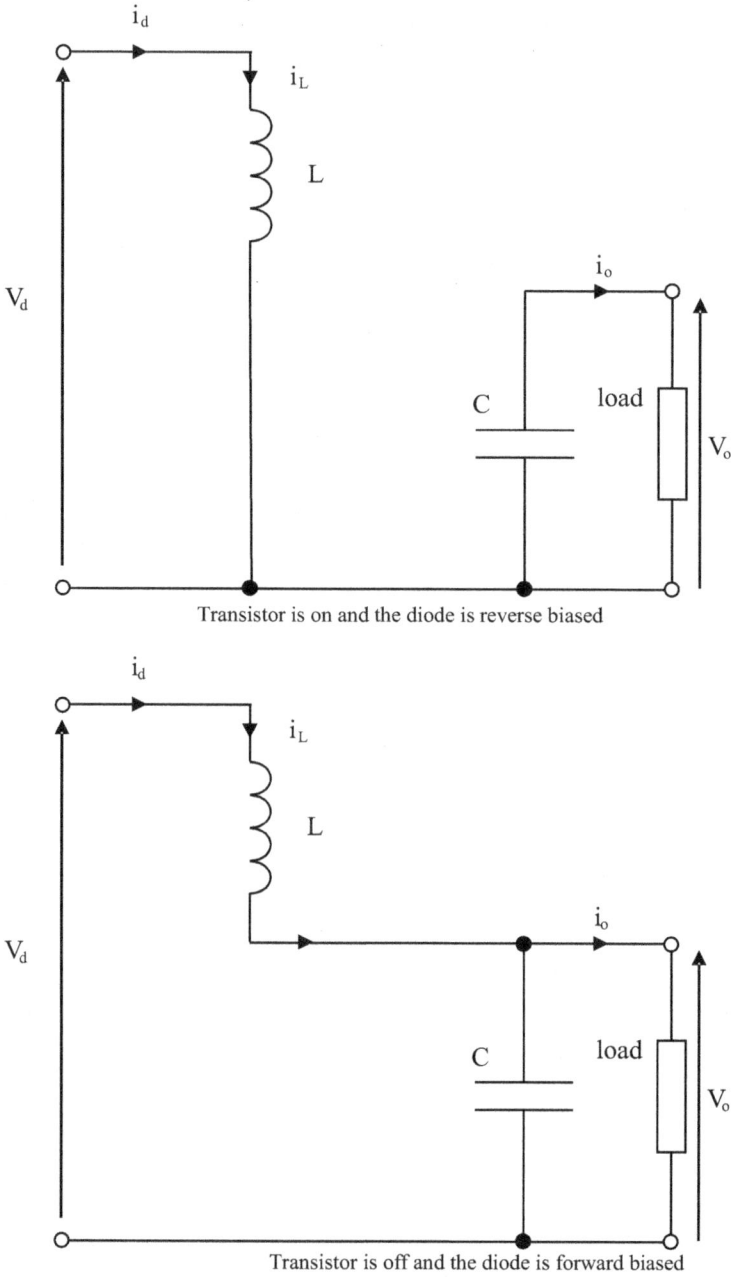

Transistor is on and the diode is reverse biased

Transistor is off and the diode is forward biased

**Figure 9.32**   The two states of the step-up circuit with an ideal transistor and diode.

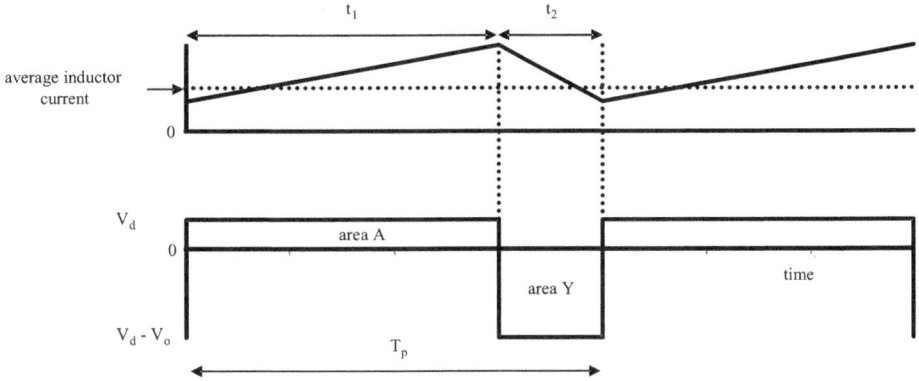

**Figure 9.33**  Waveforms for a step-up dc-to-dc converter. The inductor triangular current is shown at the top. The inductor voltage is shown at the bottom.

Therefore, the duty cycle sets the ratio between the input and output voltages. The aim of a dc-to-dc step-up converter is to maintain a constant output voltage, $V_o$, regardless of any variation in the input voltage, $V_d$. The output voltage is maintained constant by adjustments in the duty cycle.

For a triangular waveform, the average current is

$$I_{av} = \frac{I_1 + I_2}{2} \tag{9.54}$$

At the boundary between continuous and discontinuous currents, the current at the start and the end of the cycle is zero ($I_1 = 0$). The average inductor current at the boundary, $I_{Lb}$, is

$$I_{Lb} = \frac{I_2}{2} \tag{9.55}$$

During the time that the transistor is on, $t_1$

$$V_d = L\frac{I_2}{t_1} \tag{9.56}$$

Therefore,

$$I_{Lb} = \frac{t_1}{2L} V_d \tag{9.57}$$

or

$$I_{Lb} = \frac{\delta T_p}{2L} V_d \tag{9.58}$$

and

$$I_{Lb} = \frac{\delta T_p}{2L}(1-\delta)V_o \tag{9.59}$$

For no power loss in the converter, the input and output powers are equal. The average input current and inductor current are also equal.

$$V_d I_d = V_o I_o = V_d I_L \tag{9.60}$$

$$I_o = I_d \frac{V_d}{V_o} = I_L \frac{V_d}{V_o} = I_L(1-\delta) \tag{9.61}$$

The average output current at the boundary, $I_{ob}$, is

$$I_{ob} = \frac{\delta T_p}{2L}(1-\delta)^2 V_o \tag{9.62}$$

Alternatively, because the boundary output current only flows during the off time of the transistor

$$I_{ob} = \frac{I_2}{2}(1-\delta) \tag{9.63}$$

During the on time of the transistor,

$$V_d = \frac{L I_2}{\delta T_p} \tag{9.64}$$

Rearranging this equation

$$I_2 = \frac{\delta T_p V_d}{L} \tag{9.65}$$

Eliminating $I_2$

$$I_{ob} = \frac{\delta T_p V_d}{2L}(1-\delta) \tag{9.66}$$

Substituting for $V_d$

$$I_{ob} = \frac{\delta T_p}{2L}(1-\delta)^2 V_o \tag{9.67}$$

Figure 9.34 shows the characteristics of the currents and input voltage as a function of the duty cycle. The current curves peak at duty cycles of one-half and one-third for the boundary inductor and output currents, respectively. Those for the currents and output voltage with constant input voltage are shown in Figure 9.35.

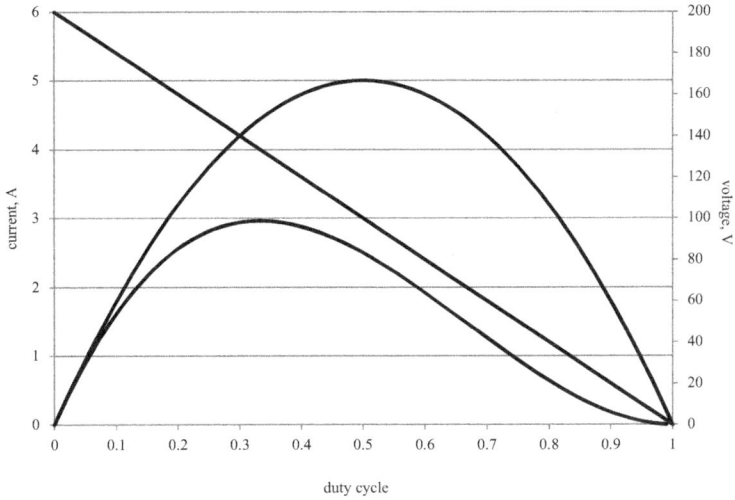

**Figure 9.34**   Inductor current (top curve), output current (lower curve) and linear input voltage (vertical axis on the right side) as a function of duty cycle for a 50-kHz and constant 200-V output dc-dc step-up converter with a 20-μH inductor.

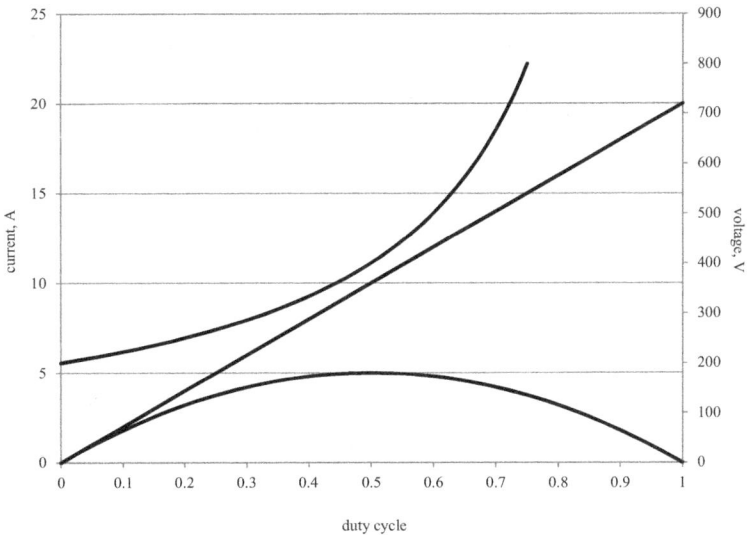

**Figure 9.35**   Linear inductor current, output current (lower curve) and output voltage (top curve) as a function of duty cycle for a 50-kHz and 200-V input dc-dc step-up converter with a 20-μH inductor.

Under discontinuous conditions, the output current falls to zero in a time of $\gamma T_p$. It is then zero for a time $(1 - \delta - \gamma)\, T_p$. As the current is steady, then

$$V_d \delta T_p = -(V_d - V_o)\, \gamma T_p \tag{9.68}$$

$$\frac{V_o}{V_d} = \frac{(\delta + \gamma)}{\gamma} \tag{9.69}$$

The peak current is

$$I_2 = \frac{V_d \delta T_p}{L} \tag{9.70}$$

$$I_2 = \frac{-(V_d - V_o)\, \gamma\, T_p}{L} \tag{9.71}$$

$$\frac{V_d \delta T_p}{L} = \frac{-(V_d - V_o)\, \gamma T_p}{L} \tag{9.72}$$

$$\gamma = \delta \frac{V_d}{(V_o - V_d)} \tag{9.73}$$

The average output current, $I_L$, which is also the average input current, $I_d$, is

$$I_L = \frac{I_2(\delta T_p) + I_2(\gamma T_p)}{2T_p} \tag{9.74}$$

$$I_L = \frac{I_2(\gamma + \delta)}{2} \tag{9.75}$$

$$I_L = \frac{V_d \delta T_p\, (\gamma + \delta)}{2L} \tag{9.76}$$

However, from the expression relating $\gamma$ and $\delta$ to the voltages

$$\gamma + \delta = \delta \frac{V_d}{(V_o - V_d)} + \delta \tag{9.77}$$

$$\gamma + \delta = \delta \frac{V_o}{(V_o - V_d)} \tag{9.78}$$

$$I_L = \frac{V_d \delta T_p}{2L} \delta \frac{V_o}{(V_o - V_d)} \tag{9.79}$$

$$I_L = \frac{\delta^2 T_p V_o V_d}{2L\, (V_o - V_d)} \tag{9.80}$$

$$\delta = \sqrt{\frac{2L\,(V_o - V_d)\,I_L}{T_p\,V_o\,V_d}} \qquad (9.81)$$

the average output current, $I_o$, is

$$I_o = \frac{\gamma I_2}{2} \qquad (9.82)$$

Substituting for $\gamma$

$$I_o = \frac{I_2 \delta V_d}{2\,(V_o - V_d)} \qquad (9.83)$$

but

$$I_2 = \frac{V_d \delta T_p}{L} \qquad (9.84)$$

$$I_o = \frac{\delta V_d T_p}{L} \cdot \frac{\delta V_d}{2(V_d - V_o)} \qquad (9.85)$$

$$I_o = \frac{\delta^2 I_2 V_d^2}{2L(V_o - V_d)} \qquad (9.86)$$

Rearranging

$$\delta = \frac{1}{V_d}\sqrt{\frac{2L I_o\,(V_o - V_d)}{T_p}} \qquad (9.87)$$

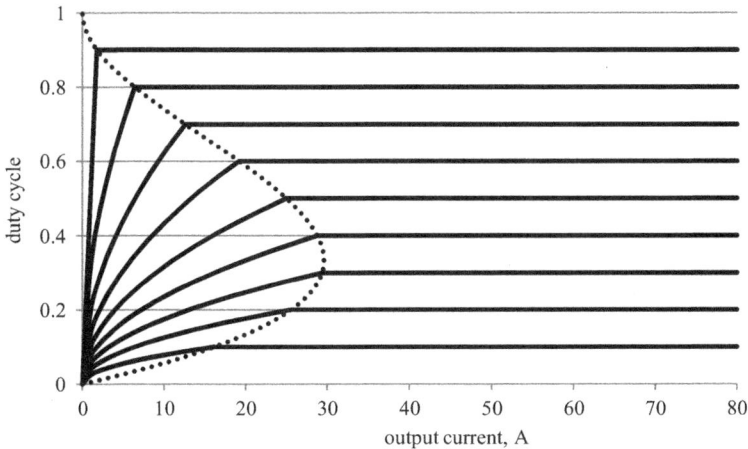

**Figure 9.36**  Continuous current on the right and discontinuous current on the left with the boundary (dotted curve). The ratios of input to output voltages ($V_o/V_d$) range from 0.9 to 0.1 in steps of 0.1 (top to bottom) ($L = 5$ μH, $f = 50$ kHz, $V_o = 100$V, $V_d$ ranged from 10V to 90V).

The two equations for the duty cycle and for the currents are related by the power equation

$$V_d I_d = V_o I_o = V_d I_L \tag{9.88}$$

Figure 9.36 shows the duty cycle as a function of the output current for ratios of input to output voltages.

## Selected Bibliography

Hart, D. W., *Introduction to Power Electronics*, Upper Saddle River, NJ: Prentice Hall, 1997.

Lander, C. W., *Power Electronics*, 3rd ed., New York: McGraw-Hill, 1993.

Mohan, N., T. M. Undeland, and W. P. Robins, *Power Electronics: Converters, Applications, and Design*, 3rd ed., New York: John Wiley & Sons, 2003.

Rashid, M. H., *Power Electronics Handbook: Devices, Circuits, and Applications*, 3rd ed., Boston, MA: Elsevier, 2011.

# 10

# Systems and Methods

## 10.1 Logic Switching

It is important that an electronic controller operates successfully in a noisy environment. Power device switching should occur smoothly with no hesitation and at the intended correct times. Otherwise, a power converter will either be destroyed in the worst case or have intermittent surges of power where overcurrent has occurred. Naturally commutated converters can recover from the occasional misfiring of thyristors. The interference can come from the power converter itself. Switching high currents and voltages in a power converter will generate electromagnetic interference from cables and power electronic components that can then be picked up by the controller. Transients induced in electronic supplies and components can be airborne or from mains supply cables.

A particular problem arises when a logic circuit is switching (Figure 10.1). If a signal is ramping up from zero to the supply voltage (second waveform from the top in Figure 10.1) and is corrupted with noise (first waveform), then switching at the midpoint of the integrated circuit supply results in multiple logic changes (third waveform) or feathering of the edge.

To overcome this problem, the use of hysteresis shown in Figure 10.2 (Schmitt trigger) results in a clean transition (fourth waveform). For an increasing signal (lower left), the switching point is at the upper threshold. As soon as the input signal becomes larger than this threshold, the output switches to the high logic level. The output will only go low again when the signal becomes less than the lower threshold.

Further precautions can be made for reliable operation. A high dc supply voltage such as 10V is preferable to a lower one. Using decoupling capacitors close to the supply pins of an integrated circuit removes high frequency pickup,

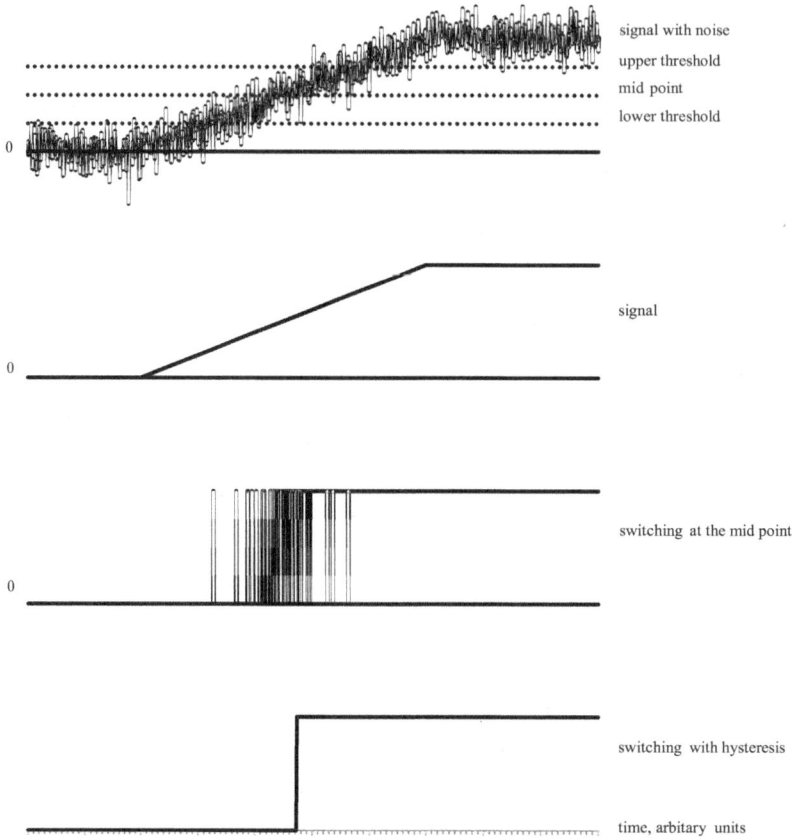

signal with noise

upper threshold

mid point

lower threshold

0

signal

0

switching at the mid point

0

switching with hysteresis

time, arbitary units

**Figure 10.1**    Logic switching showing the effects of noise and use of a Schmitt trigger to elimi-
nate multiple switching at the transition edge.

especially if they are 100 nF. Slowly rising and falling logic circuits do not
respond to unwanted high-frequency noise compared to those that have a very
high operating speed.

## 10.2  Defibrillator and Transcranial Magnetic Stimulator

In an emergency situation in which a person is having a heart attack, two con-
ductive pads are positioned across the chest and a high voltage pulse is applied.
The electrical energy causes all the heart cells to synchronize into a normal sinus
rhythm. As the electrical resistance of a person's chest varies, the resulting current is variable. A capacitor is charged and then a thyristor is fired through a
series combination of an inductor and the chest resistance. There are several sizes

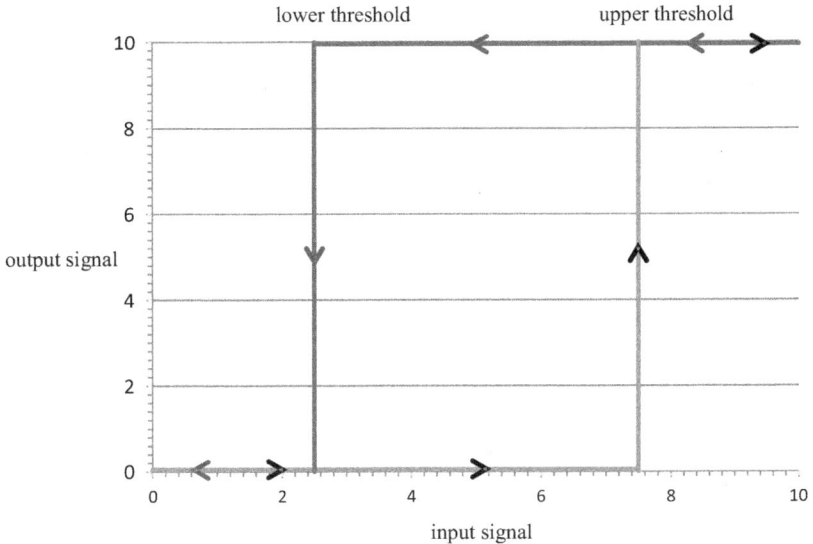

**Figure 10.2** Input and output signal characteristic showing hysteresis.

of capacitor to select the amount of energy delivered to the chest (Table 10.1). A defibrillator circuit has a second-order response. On firing the thyristor the initial current is simply the voltage on the capacitor divided by the chest resistance. The current will form a pulse, as when it falls below the holding current, the thyristor turns off (Figure 10.3).

The equation governing this circuit is given by

$$\frac{1}{C}\int_{-\infty}^{0} i\, dt + \frac{1}{C}\int_{0}^{t} i\, dt + Ri + L\frac{di}{dt} = 0 \tag{10.1}$$

Differentiating this equation

$$\frac{i}{C} + R\frac{di}{dt} + L\frac{d^2 i}{dt^2} = 0 \tag{10.2}$$

**Table 10.1**
Typical parameters for a defibrillator

| voltage | 2,500V | power | 125W |
|---|---|---|---|
| current | 50A | inductance | 40mH |
| resistance | 50$\Omega$ | capacitance | 80µF |
| stored energy | 250J (100–400) | pulse time | 4ms |

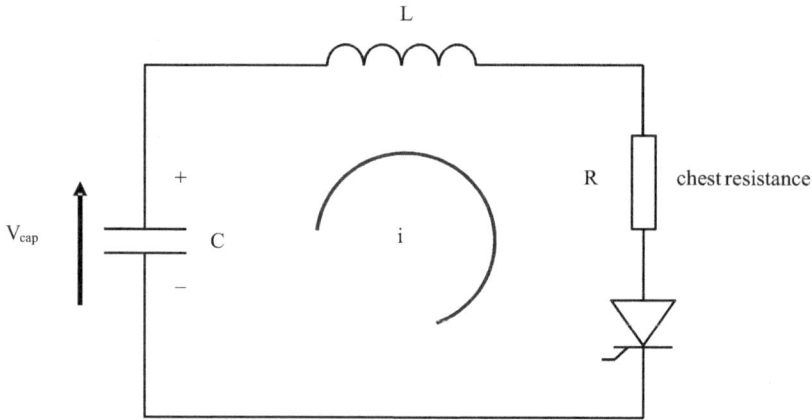

**Figure 10.3**   Circuit diagram for a defibrillator or brain magnetic stimulator.

Using Laplace transforms

$$\frac{I}{C} + RsI - Ri_o + Ls^2 I - Lsi_o - L\frac{di_o}{dt} = 0 \qquad (10.3)$$

The capacitor initially has stored energy and will produce a current, $i_o$, at time $t = 0$ (third term in the equation). The inductor has no current and so its initial current and rate of change of current are both zero (last two terms in the equation).

Rearranging (10.3),

$$I = \frac{CRi_o}{\left[1 + CRs + LCs^2\right]} \qquad (10.4)$$

There are three forms of the solution to this equation depending on the circuit parameters.

1.  Overdamped response, when the roots of the characteristic equation (setting the denominator to zero and finding the roots) are real and equal.

$$I = \frac{CRi_o}{(1 + T_1 s)(1 + T_2 s)} \qquad (10.5)$$

The transformation to time for this equation is

$$i = \frac{CRi_o}{T_1 - T_2}\left( e^{-\frac{t}{T_1}} - e^{-\frac{t}{T_2}} \right) \qquad (10.6)$$

where

$$T_1 = \frac{CR \pm \sqrt{C^2 R^2 - 4LC}}{2} \tag{10.7}$$

$$T_2 = CR - \frac{CR \pm \sqrt{C^2 R^2 - 4LC}}{2} \tag{10.8}$$

2. Critically damped response, when the roots of the characteristic equation are real and equal.

$$I = \frac{CRi_o}{(1 + Ts)^2} \tag{10.9}$$

The transformation to time for this equation is

$$i = CRi_o \frac{t e^{-\frac{t}{T}}}{T_\omega^2} \tag{10.10}$$

where

$$T_\omega = \sqrt{LC} \tag{10.11}$$

3. Underdamped response, when the roots of the characteristic equation are complex conjugates.

$$I = \frac{CRi_o}{1 + \left[ \dfrac{2\zeta s}{w_1} + \dfrac{s^2}{\omega_1^2} \right]} \tag{10.12}$$

The transformation to time for this equation is

$$i = \frac{CRi_o w_1}{\sqrt{1 - \zeta^2}} e^{-\zeta \omega_1 t} \sin w_1 \sqrt{1 - \zeta^2} t \tag{10.13}$$

where

$$\omega_1 = \sqrt{\frac{1}{LC}} \tag{10.14}$$

and

$$\zeta = \frac{R}{2} \sqrt{\frac{C}{L}} \tag{10.15}$$

A typical current waveform for an underdamped response is shown in Figure 10.4.

However, the waveform pulses for the three cases are similar. As the peak current depends on the chest resistance, a person with a small resistance will have a high current. A small capacitance can be selected first and then increased should the heart continue to fibrillate. This monophasic pulse has been replaced by a biphasic or bipolar pulse. The delivery of both positive and negative components and using different waveform shapes has led to a more efficient machine where less current is passed through the chest. In the design of a defibrillator, a power switching device is required that can withstand several kilovolts and pulses of current of about 100A. The charging circuit is designed to reach its working voltage in a short time while not stressing the capacitor with high peak currents.

A similar circuit design is used for a transcranial magnetic stimulator (TMS). A capacitor is charged and then discharged through and the inductor with the circuit resistance is kept as small as possible. Creating a large change in the magnetic field around a coil of wire positioned over the head above the motor cortex discharges neurons in the brain. In turn, the nerves in the spine send impulses to the motor alpha neurons and then to a muscle to produce a limb movement. A thyristor is a good device to allow for a single pulse of current (10,000A) passing through a figure of eight (butterfly) coil. The TMS with a low resistance has a classic resonance response with a half sine wave of current.

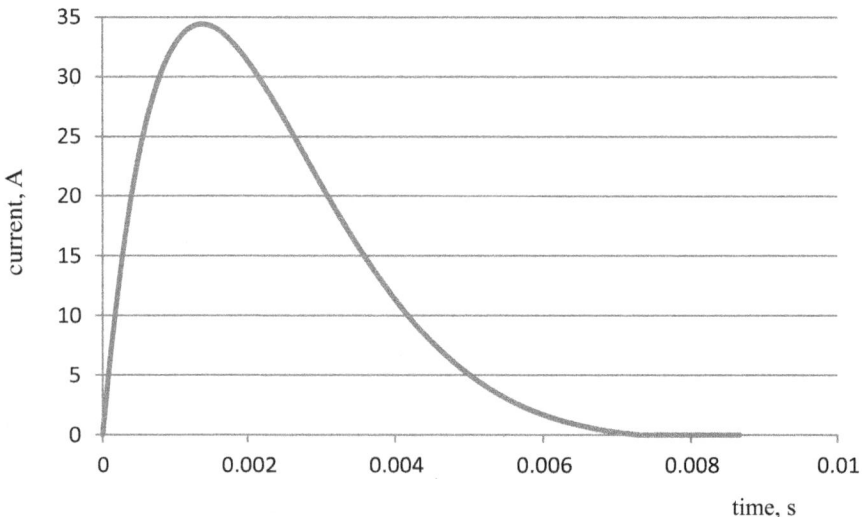

**Figure 10.4**   Under-damped response for a typical current waveform with circuit parameters of R = 50Ω, L = 39mH, C = 43μF and capacitor initial voltage of 2.5kV.

## 10.3 Synchronization to an AC Supply

The thyristors in a phase-controlled converter are fired in a sequence that is locked into the supply waveform. In a single-phase converter, a delay angle of zero is where the supply voltage is zero and is rising (zero-crossing-point). The secondary output voltage from a step-down transformer connected to the ac supply provides a low voltage and isolated signal. Electronically, the zero-crossing-point is easily found using a comparator integrated circuit or by sampling the signal digitally. However, the disadvantage of this method is that only a very small amount of the signal information is used (the voltage around the zero crossing point), that is, one sample point is used in the determination. As the ac supply contains noise and spikes, this detection method leads to uncertainty in the measurement of the zero crossing point with multiple transitions at the edge of the output waveform. An improvement is to use hysteresis (Figure 10.5). The output from the synchronization circuit is a square wave that is locked to the ac supply voltage. However, the zero crossing point has jitter due to the noise in the supply (Figure 10.6).

Figure 10.7 shows a design that uses all of the information in a sinusoidal waveform for synchronization. It is a phase-locked-loop circuit. When the circuit synchronizes to the input signal, a square wave output, at the fundamental frequency to the input signal, is produced that is locked in phase to the input. At synchronization, the input to the voltage to frequency converter is at a maximum. Alternative positive half-cycles and inverted negative half-cycles are sent to the summing amplifier via transmission gates 1 and 2, respectively. When the output is high, transmission gate 2 connects the input isolated signal to the summing amplifier. Conversely, when the output is low, transmission gate 1 connects the inverted input signal to the summing amplifier. All the half-cycles

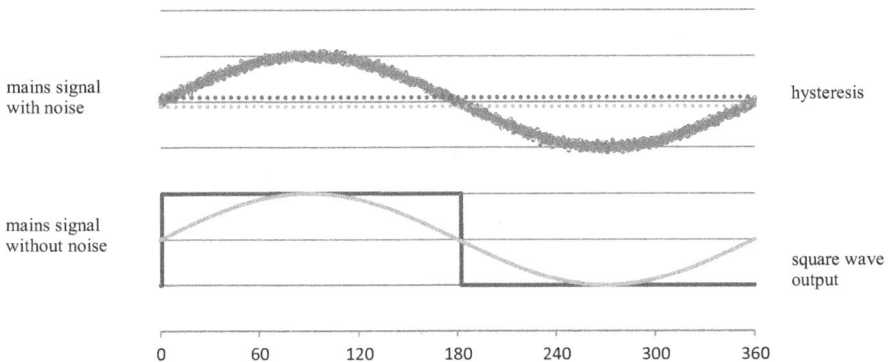

**Figure 10.5** Simple synchronization to an AC supply using a comparator with hysteresis.

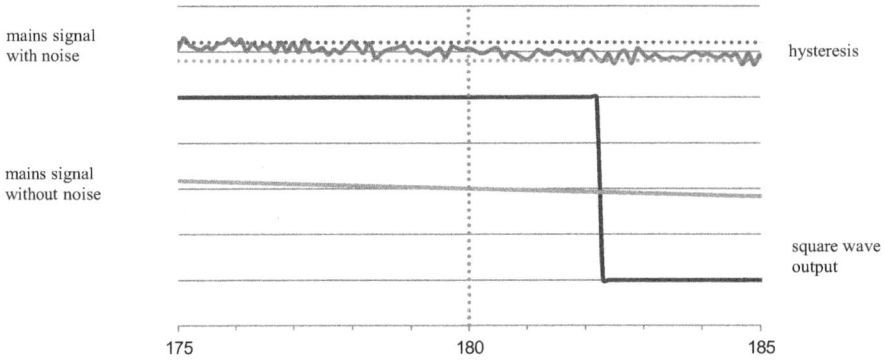

**Figure 10.6**   Expanded view of the zero-crossing-point showing edge jitter of about 1.5° and detection at 181.5° and not 180°.

of the ac supply signal at the output of the summing amplifier are positive and then smoothed by a lowpass filter or integrator. The output from the filter is a steady dc voltage and is connected to a voltage-to-frequency converter. The square wave output of the converter switches the information from the ac supply signal through the transmission gates in successive half-cycles. If the circuit is not in synchronism with the input, ac components of the input signal are averaged by the lowpass filter, resulting in a zero signal being fed to the input of the voltage to frequency converter. This feedback loop proves the synchronization technique that locks the input ac supply signal. The rising or falling edge of the output provides a reference point from which to time angular delays for phase controlled converters. The voltage-to-frequency converter has a linear characteristic, and, as the input voltage increases, so does the output square wave frequency. At its lowest setting, the output frequency should be less than the

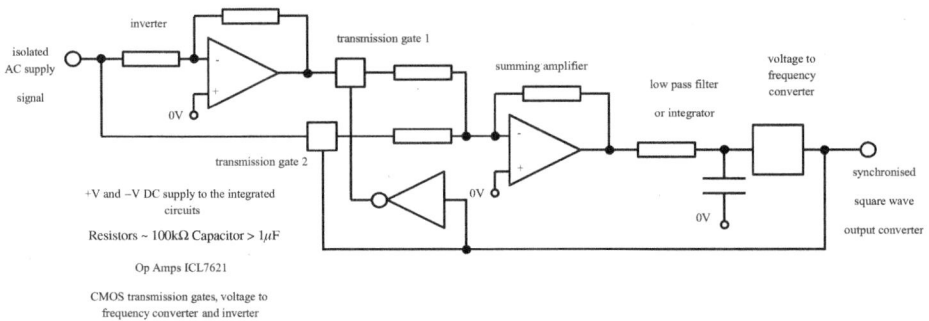

**Figure 10.7**   Synchronization circuit for a single phase supply.

expected lowest supply frequency, while at its highest, it should be above the highest frequency expected. In the United States the supply frequency varies approximately between 59.95 and 60.05 Hz. In the United Kingdom, the current regulation of the supply frequency is 50 ± 0.5 Hz. Typically, the regulation of the supply will be much better than ±10%.

For a three-phase supply, all three of the waveforms can be summed to provide the feedback signal to the voltage controlled oscillator (Figure 10.8).

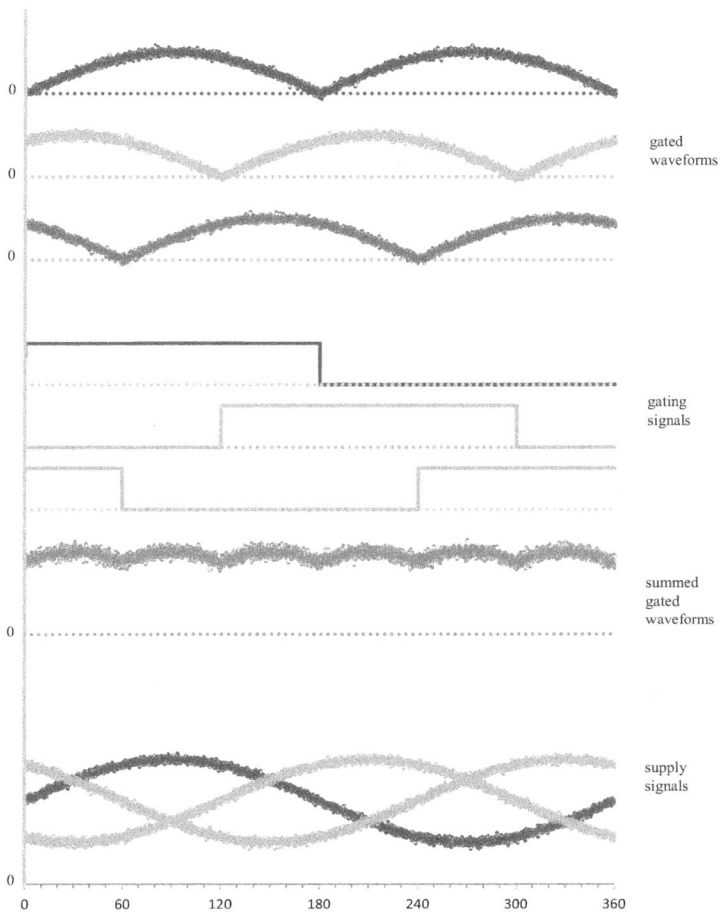

**Figure 10.8**  Synchronization to three supply signals.

When the output is locked to the supply signals, the gated logic signals are phase displaced by 120°. If the synchronization is 180° out of phase, then the inverse of the supply waveforms are summed (Figure 10.9). In practice, this situation will not happen as the input to the voltage controlled oscillator is at a low value and low frequency. If the frequency of the oscillator is different to the supply frequency, then components of the supply signals are summed (Figure 10.10). The oscillator frequency and phase adjusts automatically until it is synchronized to the supply signals (Figure 10.8).

The circuits shown in Figures 10.11 and 10.12 have three channels that are gated by three transmission gates to pass either a supply signal or its inverse to the summing amplifier. The gating signals are determined from a divide-by-6

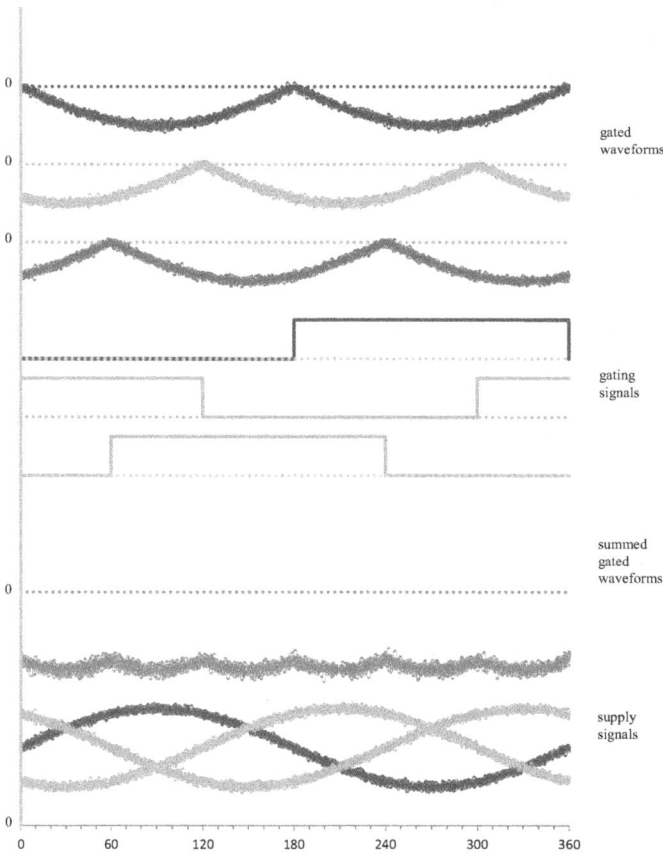

**Figure 10.9**  Synchronization that is 180° out of phase showing a negative summed waveform.

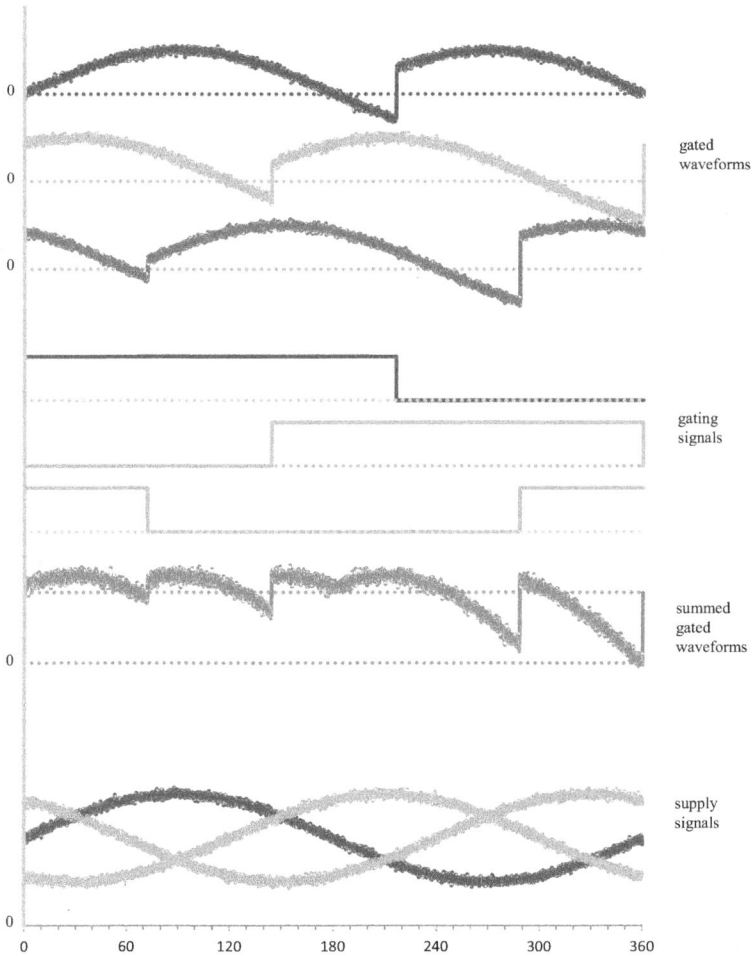

**Figure 10.10**  Gating signals operating at 20% frequency than the supply signals. The digital signals are not synchronized to the supply and the average gated signal is lower than that when there is synchronism.

counter that is clocked by the oscillator at 360 Hz for a 60 Hz supply or 300 Hz for a 50-Hz supply. Shown in Figure 10.13 are the digital waveforms from the counter.

The clock used to drive the divide-by-6 counter can be further divided to provide the finer resolution needed for the phase control of a thyristor converter. A minimum resolution of 0.5° requires a clock operating at 36 kHz for a 50-Hz supply (43.2 kHz for a 60-Hz supply). A divide-by-120 counter provides this resolution. Alternatively, the divide-by-6 counter in Figure 10.13 can be

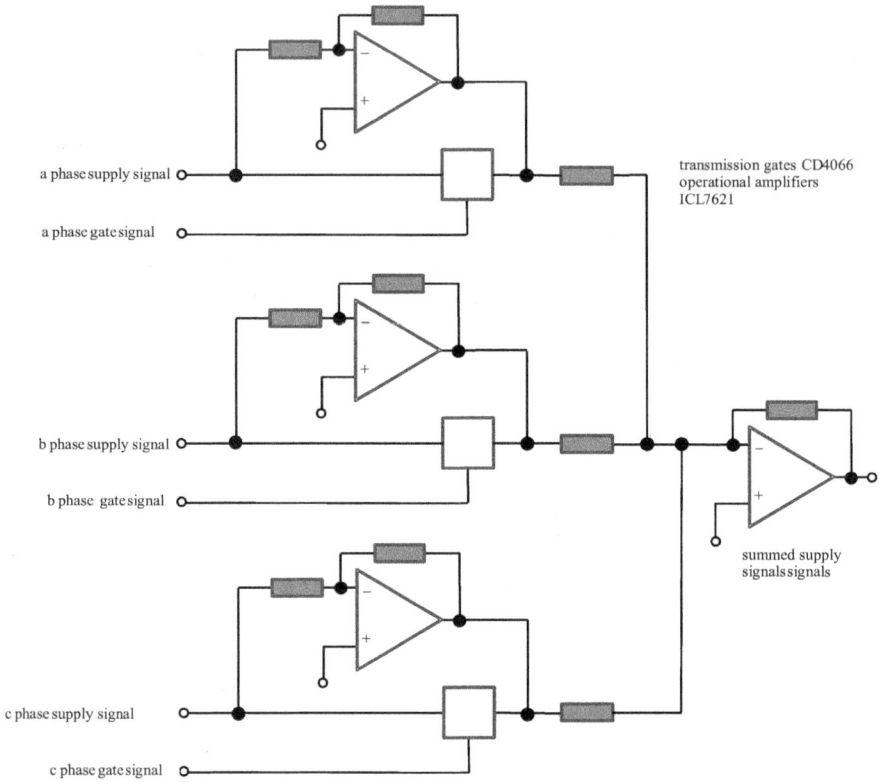

a phase supply signal

a phase gate signal

transmission gates CD4066
operational amplifiers
ICL7621

b phase supply signal

b phase gate signal

summed supply
signals signals

c phase supply signal

c phase gate signal

**Figure 10.11**   Three-phase summation circuit for the noninverted and inverted signals.

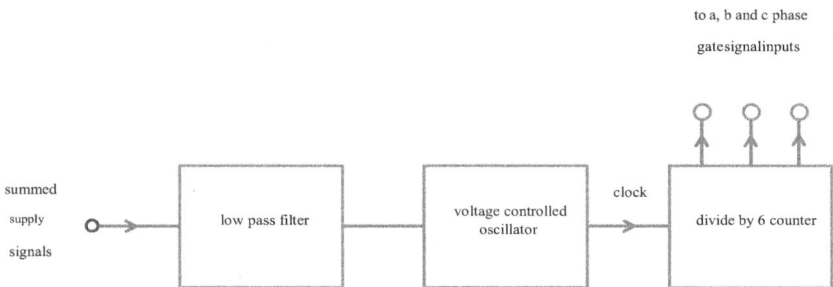

to a, b and c phase

gate signal inputs

summed

supply

signals

low pass filter

voltage controlled
oscillator

clock

divide by 6 counter

**Figure 10.12**   Low pass filter, voltage controlled oscillator and decoding for the three gate
signals.

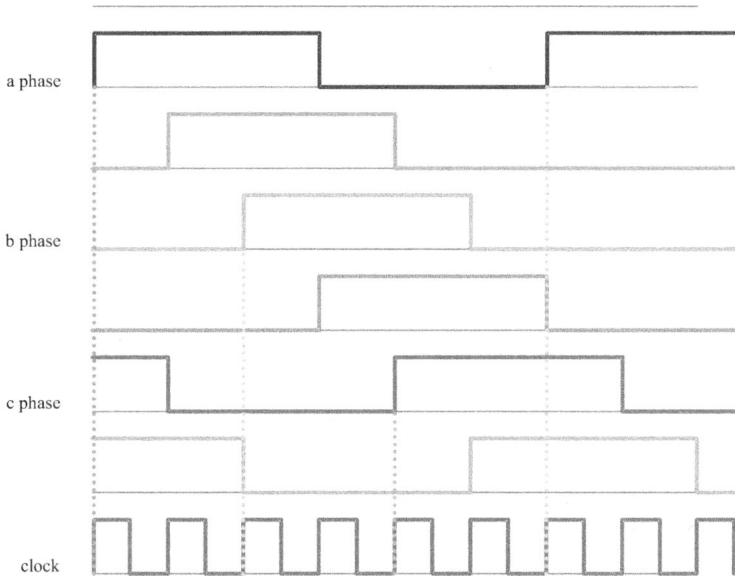

**Figure 10.13** Divide-by-6 counter showing gating signals for the transmission gate inputs.

replaced by a divide-by-720 counter to provide the gate signals and high resolution clock together in one loop. The output of the oscillator is scaled to operate at the higher frequency in the tens of kilohertz.

Integrated analog and logic circuits are used in the description of the operation of synchronization to an ac supply to provide the reader will an appreciation of the elements that form a working circuit. However, all of these functions can be more conveniently assigned to a microcontroller or other programmable device. The supply signals are sampled and an algorithm is written for the synchronization. Counters and timers are used for the higher-resolution clock.

## 10.4 Measurement of Current and Voltage

A digital storage oscilloscope (DSO) is essential equipment for observing waveforms in power electronic circuits. By setting an input of a DSO to dc (rather than ac) and the trigger to dc, a single sweep of a waveform can be stored and analyzed. During prototype development, when a power converter may operate erratically, repeat captures of waveforms can be displayed. In the design of a DSO, there is a trade-off among sampling rate, analog-to-digital conversation resolution, number of channels, and number of samples captured. Most oscilloscopes have an 8-bit vertical resolution corresponding to 256 levels for each of four channels. However, with this limited resolution, the sampling rate is high and of the order of 1 GSs$^{-1}$. The number of samples per record is also good and can be increased with a more expensive oscilloscope. Circuit development and fault finding is enhanced by the pretrigger function of a DSO. The trigger specifications are set to capture the conditions of circuit malfunction. An input channel is set to its dc mode with the normal mode setting for the trigger. Further settings on the trigger are rising edge (or falling edge) and dc coupling. The trigger point is also set so that it is above the noise and unwanted transients in the signal to be captured. The trigger delay can be adjusted to observe voltages and currents just before or after the critical event. An occasional and unexpected current surge in a power converter can trigger the capture of data on one channel and three other voltages observed in the controller or converter at the same time. A further feature is to set the DSO into a data logger mode when longer-term events over a day are stored.

Most DSOs have a grounded case and inputs. Care must be taken to make sure that the outer connector (of a BNC plug and socket) is not connected to an unearthed conductor of a power circuit. A few DSOs have isolated inputs that allow for the observation of high voltages. These oscilloscopes use special high-voltage probes that can also withstand high voltages. A useful probe for electronic controller development is one that attenuates a signal by 10 times (10×). For higher voltages, attenuation by 100 (100×) or 1,000 (1,000×) are used. Other useful features of a DSO are mathematic functions such as waveform addition, subtraction, multiplication, frequency, peak-to-peak voltage, period, mean, root-mean-square, minimum, maximum, and fast Fourier transform.

A DSO can easily store samples of the supply waveforms to estimate leakage inductance in a three-phase supply. Connecting a three-phase diode bridge to a load resistor results in overlap as discussed earlier. The lost voltage-time area can be found with calculations from the stored waveforms by importing them into a spreadsheet or by writing a simple program in MATLAB®. Measurement of the resistor current will then lead to an estimate of the leakage inductances.

An alternative method to achieve isolation between the earthed inputs of a DSO and high-voltage ac signals is to use step-down transformers. These are made in a wide variety of sizes and specifications. If the voltage to be observed has high-frequency components, then a small pulse transformer can be used to good effect. Such a device has small capacitances between the primary and secondary windings. The use of analog opto-couplers also achieves isolation of several kilovolts and transmission of high-frequency signals.

Care must be taken not to expose anyone to dangerous voltages or hot components. A common precaution is to only use only one hand when grasping or touching equipment or attaching an oscilloscope probe, especially if the power converter is a prototype and therefore under development. The other hand can be put out of the way and is often put in a pocket. This technique avoids a dangerous potential difference developing between the hands and directly across the chest that could transmit a fibrillating current through the heart. It is still possible for a dangerous potential difference to occur between one hand and another body part that is at or near an earthed conductor. However this precaution is not a substitute for making sure that all live and potentially live parts and conductors in a converter are insulated and isolated from people. If there are exposed cables or metalwork, all of these should be earthed. In laboratories, while developing a power converter, it is tempting sometimes to save time and not shut off the power supply before making adjustments to the converter. This carelessness can lead to hazardous situation since live connectors and cables can become exposed. Also any stored energy in passive components, especially inductors, can lead to transient high voltages and sparks. When a power converter is not in use, dangerous voltages will remain on capacitors unless they are discharged. The stored energy can be dissipated if a resistor is connected in parallel with the capacitor before the converter is connected to a power supply. Thus, ensuring the voltage will exponentially decay to zero when the convert is not in use. It is always better to use the control electronics to bring the converter down to a safe and low energy state before switching off the power supply, disconnecting the converter, and then making any changes to the power converter components or layout. The experimental development of power electronics requires different and more cautious methods compared to the development of low voltage (less than 30V) electronic circuits.

Observation of current in cables and components requires a transducer. A common sensor is based on the Hall effect. These transducers have a magnetic core with a semiconductor Hall effect device sliced into the core (Figure 10.14). Some varieties allow for the transducer to be clipped into a circuit without the

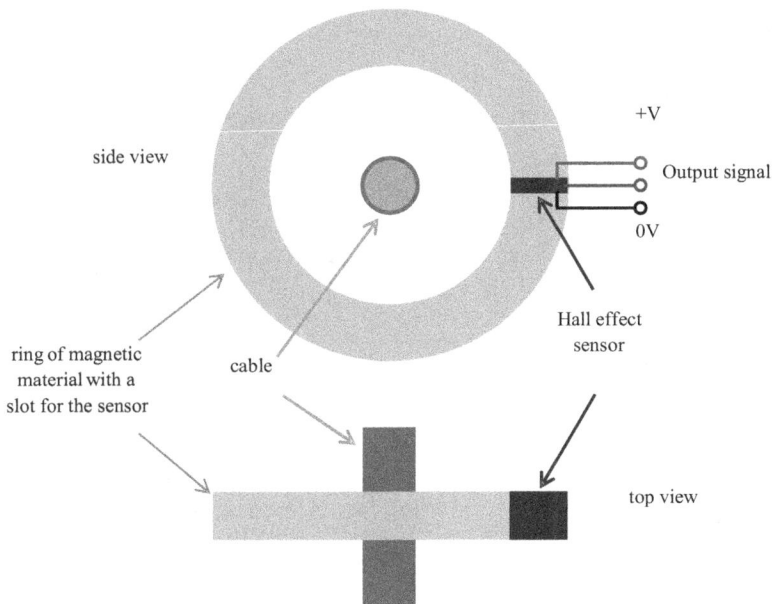

**Figure 10.14**    Schematic diagram of a Hall effect current transducer.

need to disassemble connecting wires. Looping a wire several times through the central core increases the sensor sensitivity.

Another method is to measure the voltage across a resistor with an instrumentation amplifier (Figure 10.15). In this circuit, the resistance of the sensor is chosen to be a small value. The resistor will then dissipate a small amount of power and there will be a small voltage across its terminals even when there is the maximum current. By measuring only a few tens or hundreds of millivolts, the circuit operation is not disrupted. A limitation of this circuit is that, unlike the Hall effect sensor, there is no electrical isolation from the power converter voltages. Isolation can be achieved using optical components. Special resistors are designed to be used in power circuits and are called current shunts. These resistors have a high specification to measure current in the range from less than an ampere to several tens of amperes. They can be connected to a DSO for waveform observation with suitable isolation.

## 10.5 Lowpass Filter

A simple circuit consisting of a series connected resistor and capacitor has many uses in power electronics. It can be used to create a signal delay in electronic controllers or as a filter in a power converter. Its gain and phase characteristics are shown in Figure 10.16.

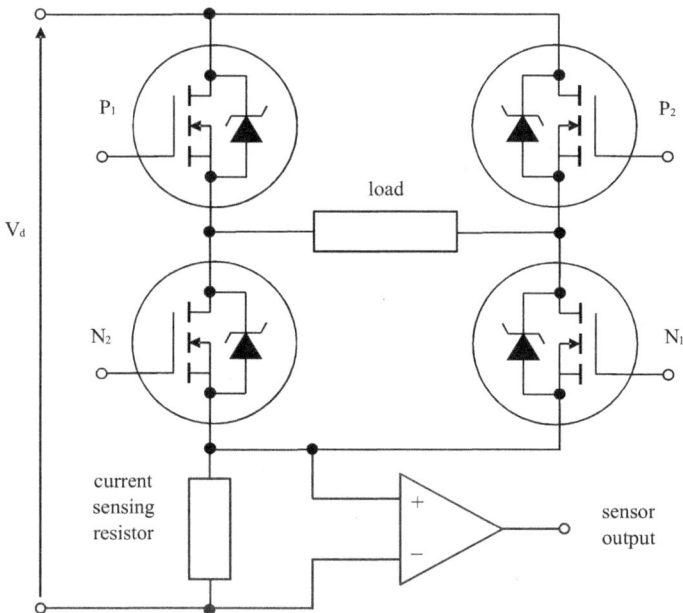

**Figure 10.15**  Example of sensing current with a resistor in an H-bridge circuit.

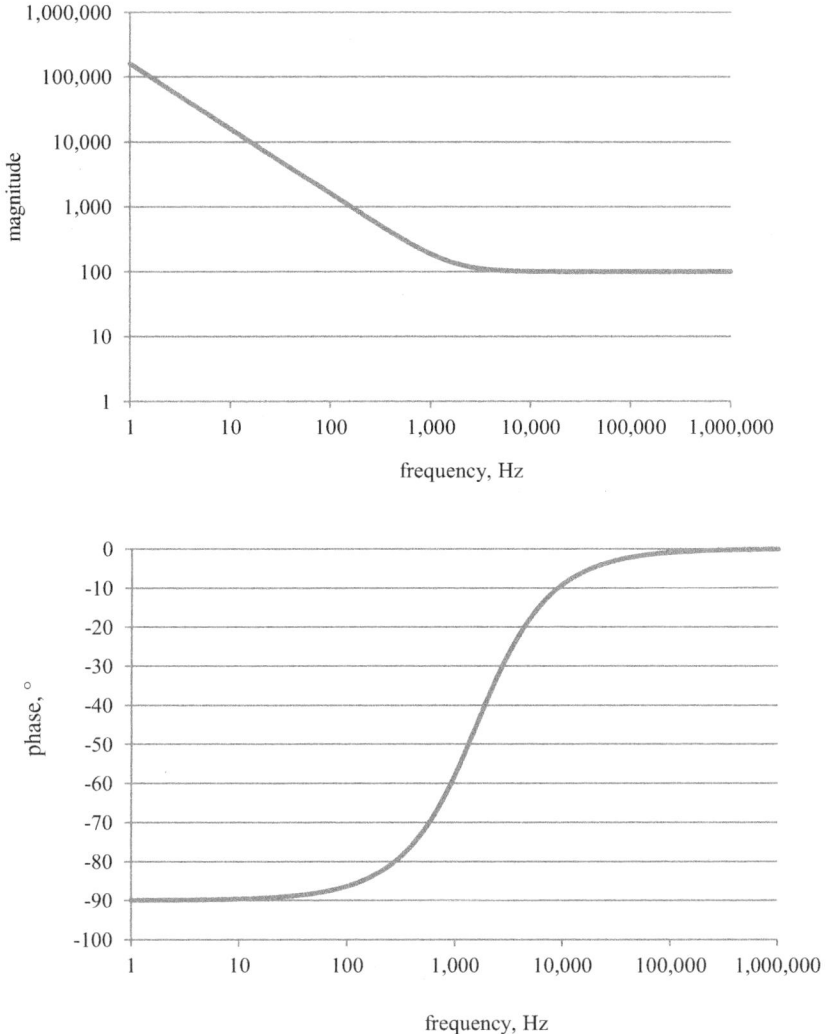

**Figure 10.16** Characteristic of a lowpass filter for a series resistor (10Ω) and capacitor (1 nF).

One of most useful functions of this circuit is in the design of a snubber, which can include a resistor and a diode to limit the rate of change of voltage across the terminals of a power device.

## 10.6 Circuit Protection

During the development of a power converter, it is useful to measure the supply current and, when a fault occurs, to disconnect the converter rapidly from the supply. Quick blowing fuses provide circuit protection as well, but it is more convenient to reset the supply contactor set and avoid the cost of fuse replacement.

Figure 10.17 shows a schematic diagram for a three-phase system. The three line currents are monitored by current sensors (top half of the diagram)

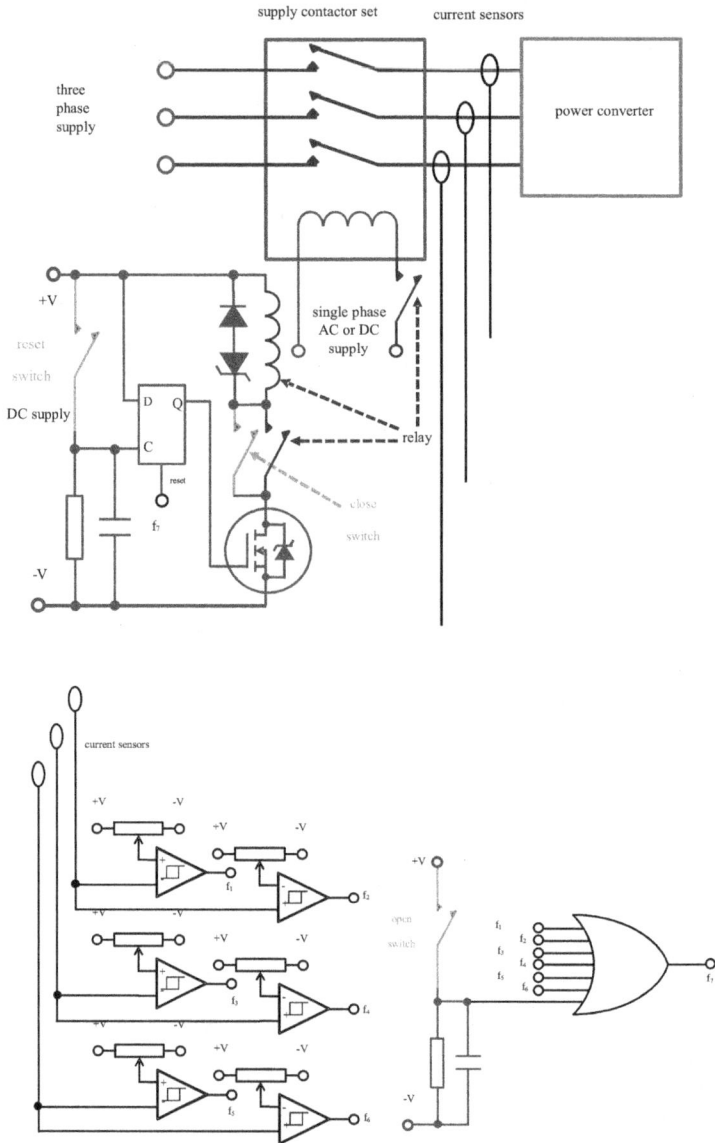

**Figure 10.17**　Three-phase contactors with over current measurement and protection.

and their signals sent to comparators and logic control (lower half of the diagram). To start the system, the output (Q) of a D-type flip-flop is set to a logic one (+V) by momentarily closing the reset switch. This action generates a rising edge that clocks (C) the one at the input (D) to the flip-flop through to the output (Q). The MOSFET transistor is then turned on. Momentarily closing the switch connected to the drain of the MOSFET activates a relay coil. This action causes the relay contactors connected to the supply contactor set to close. The power converter is then connected to the three-phase supply. A second contactor in the relay maintains power through its coil as the closing switch is open circuit. This arrangement provides an interlock.

If the line current peaks above a preset threshold, then the supply contactors are opened. Six comparators with hysteresis have the threshold set by potentiometers that are connected to their noninverting input to monitor the current in positive half-cycles and to the inverting input to monitor the negative half-cycles (lower diagram in Figure 10.17). If a line current increases to a value above the set limits, then a comparator output goes high (either $f_1$, $f_2$, $f_3$, $f_4$, $f_5$, or $f_6$). The output of these signals ($f_7$) resets the D-type flip-flop to a logic zero (−V). The output goes low and the MOSFET is turned off. The current in the supply contactor set is removed and the supply contactors open. Momentarily closing the open switch connected as one of the inputs to the OR gate has the same effect as overcurrent detection.

A similar technique to that used for an ac supply is used to detect over current in a dc supply (Figure 10.18). In this circuit, there are an inductor and a thyristor connected in parallel with the power converter. On detecting an overcurrent, a pulse is sent to the thyristor gate so that the thyristor is triggered and the supply current is diverted from the power converter to the inductor while the supply contactors are disconnecting the power converter from the supply.

## 10.7 Fourier Series

Power converters output current and voltage waveforms that are applied to an electrical load. As power semiconductors are switched, harmonic components are present in an ac supply as well as the desired fundamental frequency. In a dc system the presence of ripple currents may lead to unwanted power dissipation in a load. Standard waveforms can be analysed using Fourier series while components in experimental data can be found using fast Fourier transforms. The harmonic content of a power electronic waveform can be predicted using Fourier series for a function, $f(x)$.

$$f(x) = \frac{a_0}{2} + \sum_{r=1}^{\infty} (a_r \cos rx + b_r \sin rx)$$

(10.16)

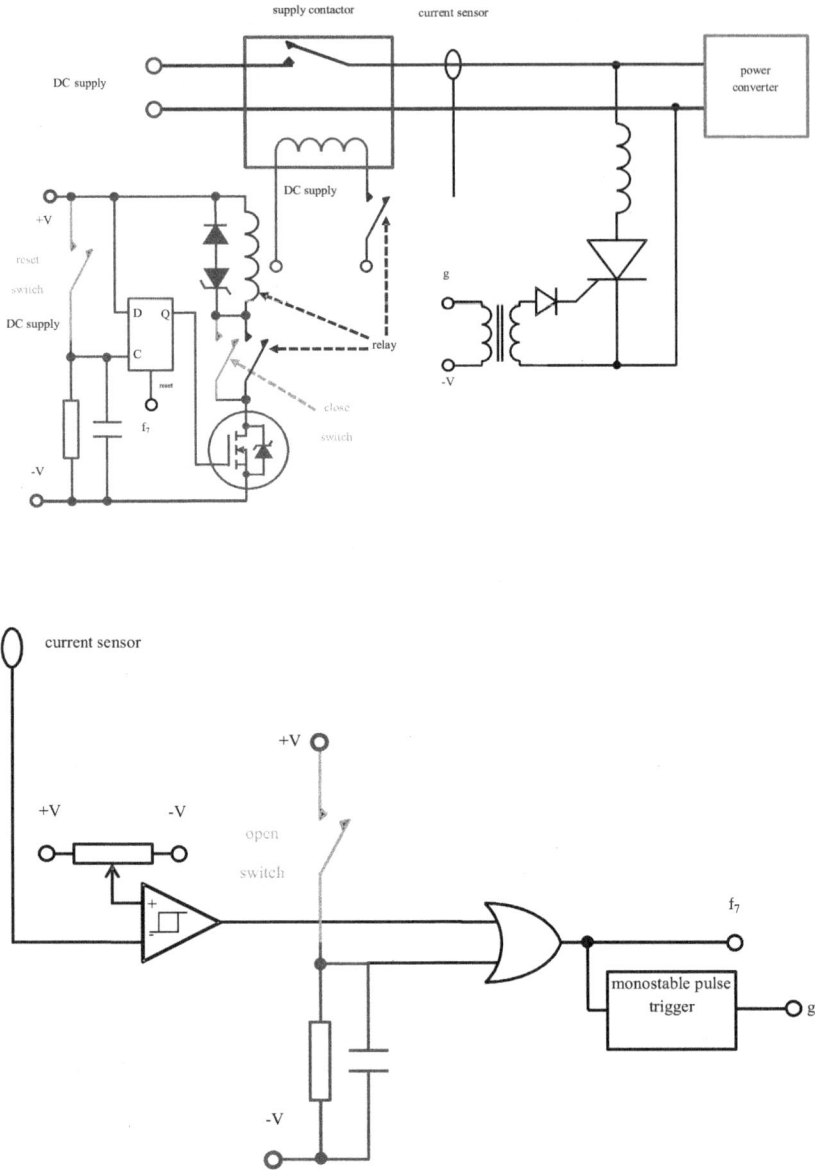

**Figure 10.18**  DC supply with over current protection and an inductor with a thyristor crowbar.

$$a_0 = \frac{1}{\pi} \int_{-\pi}^{\pi} f(x) \ dx \tag{10.17}$$

$$a_r = \frac{1}{\pi} \int_{-\pi}^{\pi} f(x) \cos rx \ dx, \quad (r = 1, 2, 3, \ldots) \tag{10.18}$$

$$b_r = \frac{1}{\pi} \int_{-\pi}^{\pi} f(x) \sin rx \ dx, \quad (r = 1, 2, 3, \ldots) \tag{10.19}$$

For example, a square wave has an odd number of cosine components.

$$f(t) = \frac{F}{2} + \frac{2F}{\pi} \left( \cos \omega_1 t - \frac{1}{3} \cos 3\omega_1 t + \frac{1}{5} \cos 5\omega_1 t - \cdots \right) \tag{10.20}$$

where $F$ is the magnitude and $\omega_1$ is the fundamental component of the waveform.

The switching in a power electronic circuit will interfere with the quality (pure sine wave with voltage and current in phase) of an ac power supply. Hence, there are various national and international standards to put limits on the harmonic currents and voltage distortion at different frequencies. These restrictions are set by countries and groups of countries, for example, the European Union.

- EN 50 006, "The Limitation of Disturbances in Electrical Supply Networks Caused by Domestic and Similar Appliances Equipped with Electronic Devices": a European standard by Comité Européen de Normalisation Electrotechnique (CENELEC).

- IEC Norm 555-3 by the International Electrical Commission West German Standards, VDE 0838 for household appliances, VDE 0160 for converters, and VDE 0712 for fluorescent lamp ballasts.

- IEEE Guide for Harmonic control and Reactive Compensation for Static Power Converters and ANSI/IEEE Std. 519-1981 and revision in 1992 to 519-1992.

As well as limits on individual components ($I_n$), the total harmonic distortion (THD) is also specified with respect to the fundamental component ($I_1$).

$$TDH = \frac{\sqrt{\sum_{n=2}^{\infty} I_n^2}}{I_1} \tag{10.21}$$

It is the ratio of the squared amplitudes of all the harmonics summed together compared to the fundamental component of the current.

## 10.8 AC Line Transient Suppression

Voltage spikes appearing in the lines of a three-phase supply can affect the operation of a power converter. There are several devices that are designed to remove transient voltages. These components have an open circuit characteristic (high resistance) but become a short circuit above a threshold voltage. A zener diode is an example of such a device. Another method is to connect the supply to a diode bridge (Figure 10.19). The capacitor will charge to the peak of the supply voltage and any spike above this value will be suppressed. The resistor acts to damp the response of the circuit and to discharge the capacitor when the supply is disconnected.

## 10.9 Efficiency of DC Converters

The supply of a dc converter is often fixed for the application. In a car the typical battery voltage is 12V, although this may increase with the demand for more automation to 24V or more to reduce the current and copper in the cables. With an increase in voltage also comes the added benefit of a fall in the losses in a power converter when considering the on-state voltage and current of a power semiconductor.

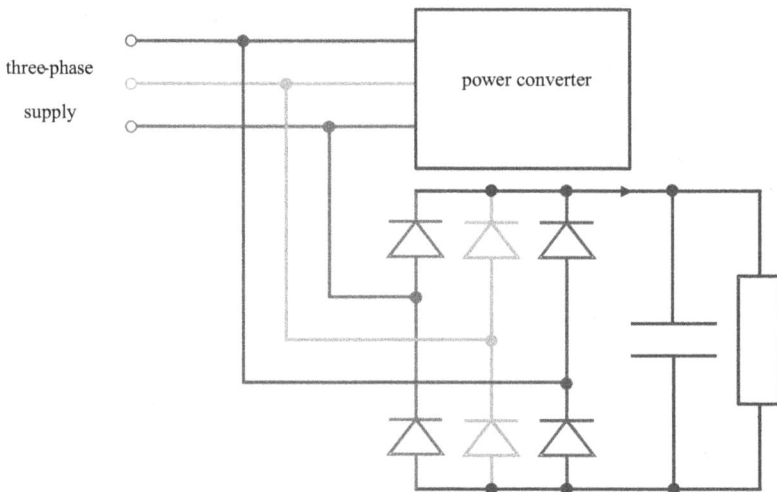

**Figure 10.19**   Transient suppression of a three-phase supply using a diode bridge, capacitor and resistor.

Consider a supply voltage, $V_d$, when two power devices are conducting current and have a combined series resistance, $R$. The series load has a voltage, $V$, across its terminals and the power in the circuit is

$$V_d = Vi + i^2 R \qquad (10.22)$$

During the time that the devices are on, the efficiency expressed as a percentage is

$$\eta = 100 \left( \frac{V_d i - i^2 R}{V_d i} \right) \qquad (10.23)$$

or

$$\eta = 100 \left( 1 - \frac{R}{R_p} \right) \qquad (10.24)$$

where

$$R_p = \frac{V_d}{i} \qquad (10.25)$$

Equations (10.23) and (10.24) are plotted in Figures 10.20 to 10.23, for values of $i$, $V_d$, $R$, and $R_p$. The lines in these plots can be used to estimate efficiency. For example, if the power devices are MOSFETs, a manufacturer will specify a value for $R_{DSon}$. The resistance, $R$, can be calculated by summing the

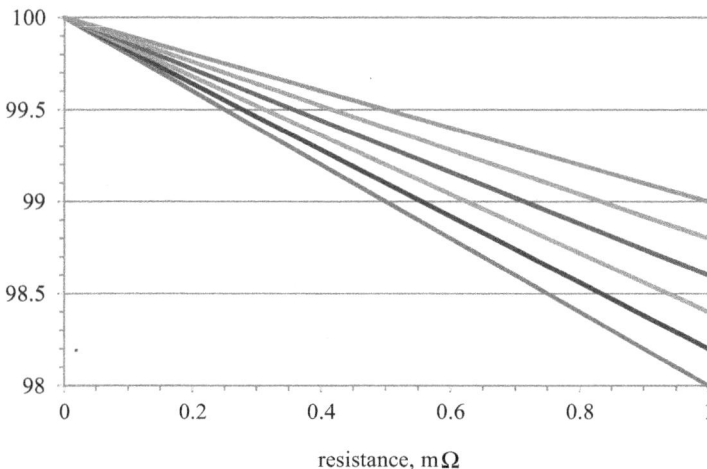

**Figure 10.20**  Efficiency of power devices as a percentage against series on-state resistance, R, for values of Rp (Vd / i) ranging from 0.1 to 1 mΩ (bottom line to the top line).

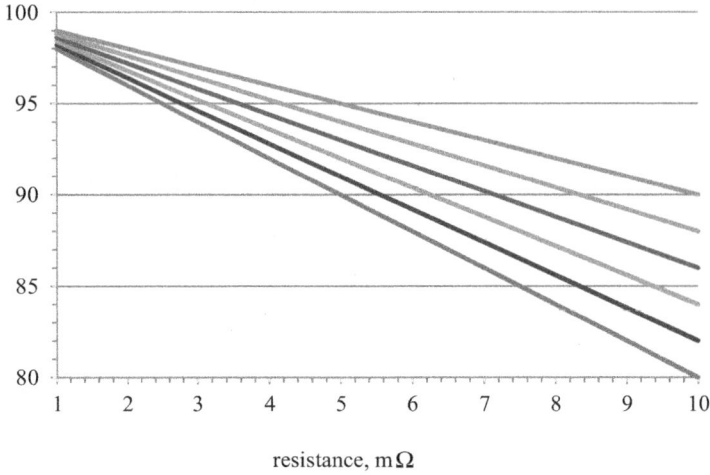

**Figure 10.21**   Efficiency of power devices as a percentage against series on-state resistance
$R$, for values of $Rp$ ($Vd/i$) ranging from 1 to 10mΩ (bottom line to the top line).

total series resistance, for the conducting devices (one or more typically two)
which forms the x-coordinate on one of the plots. The resistance, $R_p$, can be calculated from the supply voltage, $V_d$, for the given application and the predicted
maximum current through the load, $i$. A line (or the nearest line) is selected from
one of the six lines on the plot and the efficiency can be read from the y-axis.

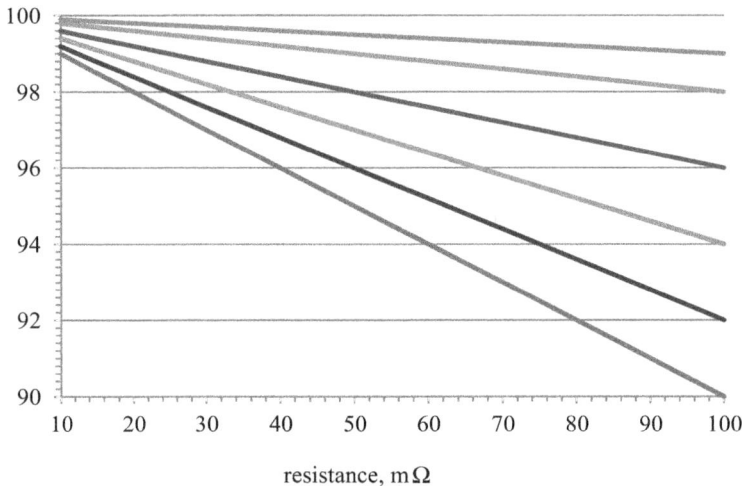

**Figure 10.22**   Efficiency of power devices as a percentage against series on-state resistance,
$R$, for values of $Rp$ ($Vd/i$) ranging from 10 to 100mΩ (bottom line to the top line).

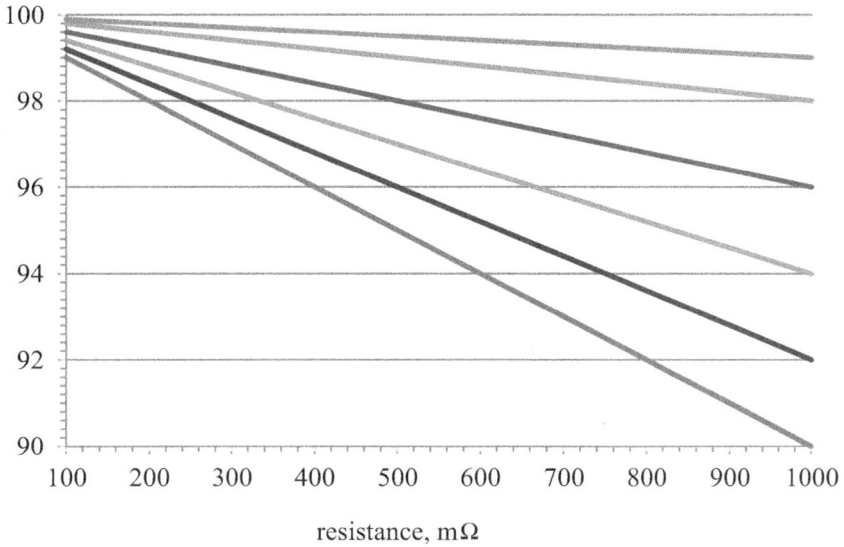

**Figure 10.23** Efficiency of power devices as a percentage against series on-state resistance, *R*, for values of *Rp* (*Vd*/*i*) ranging from 100 to 1000 mΩ (bottom line to the top line).

## 10.10 Chaotic Behavior in Diode Circuits

Under certain combinations of ac supply frequency and amplitude, a diode connected in series with an inductor (Figure 10.24) behaves chaotically. At frequencies well above the normal low supply frequencies of 50 or 60 Hz, the voltage across the inductor can have amplitudes above the supply voltage (Figure 10.25).

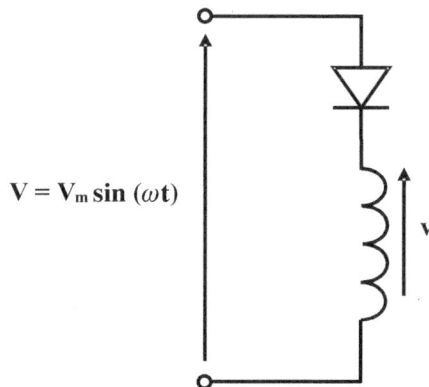

$$V = V_m \sin (\omega t)$$

**Figure 10.24** Circuit for an experimental investigation of the chaotic behavior of a diode and inductor input voltage.

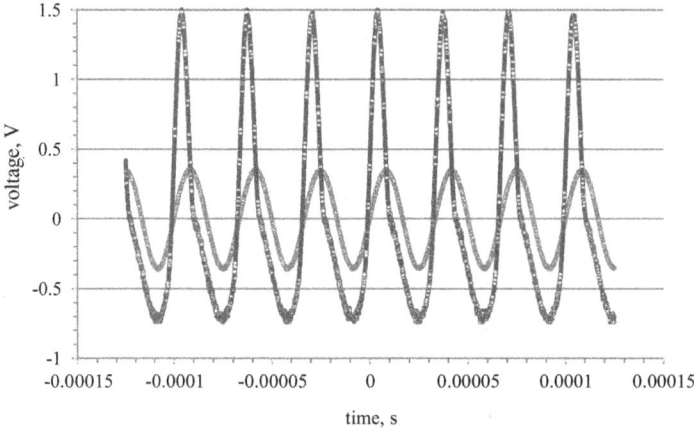

**Figure 10.25**    Increased voltages across an inductance of 25.5 mH at 30 kHz with a series VS-70HF160 diode (70A and 1600V). The supply voltage (centered on zero) is also shown in the plot.

Increasing the supply voltage causes the inductor voltage to become chaotic (Figure 10.26). Increasing the supply voltage further, the voltage peaks halve in frequency (Figure 10.27). A higher voltage results in chaotic behavior, followed by the peaks occurring at one in three of the supply frequency (Figure 10.28).

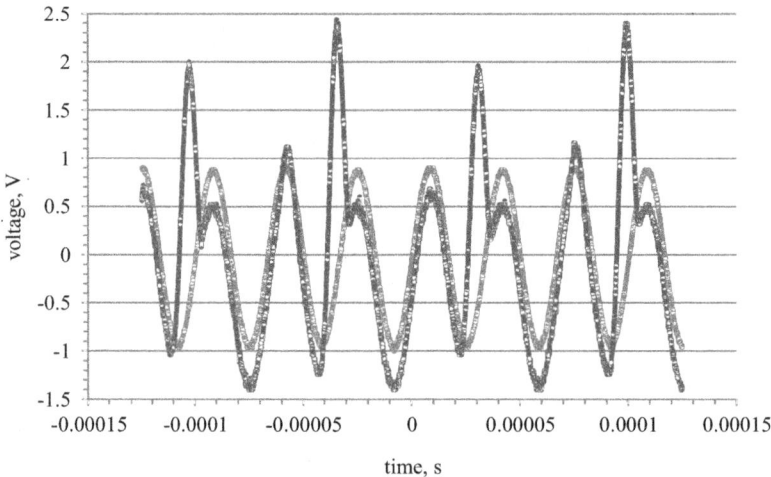

**Figure 10.26**    Chaotic behavior in the voltage across an inductance of 25.5 mH at 30 kHz with a series VS-70HF160 diode (70A and 1600V). The supply voltage (centered on zero) is also shown in the plot.

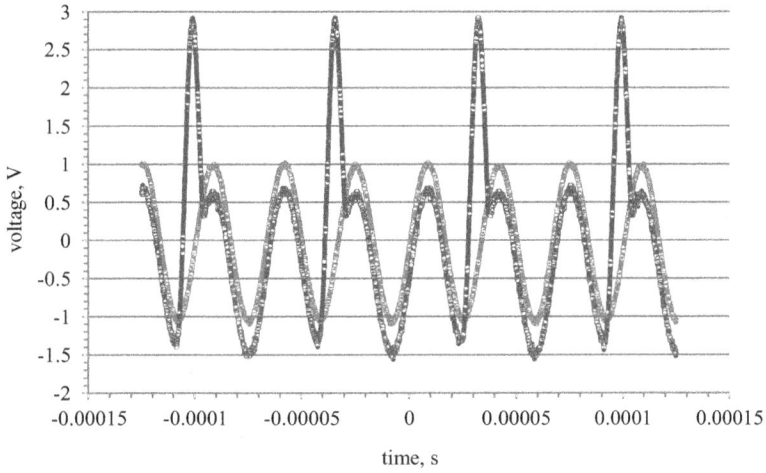

**Figure 10.27** Halving of the frequency in the voltage peaks across an inductance of 25.5 mH at 30 kHz with a series VS-70HF160 diode (70A and 1600V). The supply voltage (centered on zero) is also shown in the plot.

This process of chaotic behavior followed by a decrease in the frequency of the peaks continues with higher supply voltages. In Figure 10.29 the peaks occur at a rate of one-quarter of the supply frequency. This behavior also occurs in wave rectifiers (e.g., for a single-phase supply with four diodes).

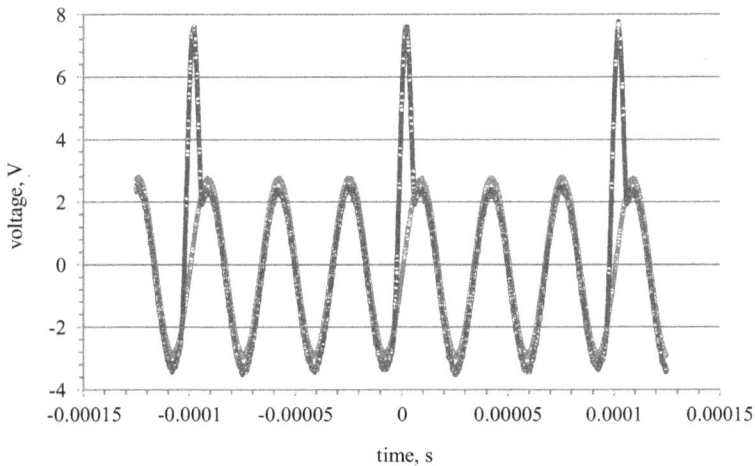

**Figure 10.28** Peaks of voltage occurring at one in three of the supply cycles across an inductance of 25.5 mH at 30 kHz with a series VS-70HF160 diode (70A and 1600V). The supply voltage (centered on zero) is also shown in the plot.

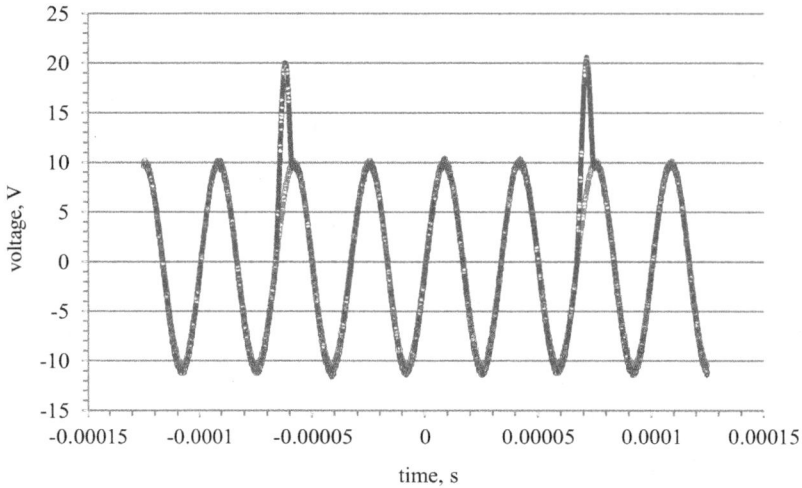

**Figure 10.29** Peaks of voltage occurring at one in four of the supply cycles across an inductance of 25.5 mH at 30 kHz with a series VS-70HF160 diode (70A and 1600V). The supply voltage (centered on zero) is also shown in the plot.

It is the nonlinear relationship of a diode that leads to bifurcation [1]. With a high Q inductor, the peaks are more pronounced. This phenomenon may account for unpredictable events in power converters, especially when operated at frequencies in the tens or hundreds of kilohertz.

## Reference

[1] Testa, J., J. Perez, and C. Jefferies, "Evidence for Chaotic Behaviour of a Driven Nonlinear Oscillator," *Physical Review Letters*, Vol. 48, No. 11, 1982, pp. 714–717.

## Selected Bibliography

Brown, B. H. et al., *Medical Physics and Biomedical Engineering*, Philadelphia, PA: IOP Publishing, 2001.

# 11

# Examples

## Example 11.1

Calculate the mean voltage for a half wave rectifier and full-wave rectifier that has a single-phase supply voltage of 120V. The anode-cathode voltage of each diode is 1.1V in forward conduction.

### Solution

$$V_{av} = \frac{1}{T} \int_0^T v(t) \, dt \tag{11.1}$$

For a half-wave rectifier, positive sinusoidal voltages occur for half the period.

$$V_{av} = \frac{1}{2\pi} \int_0^\pi V_m \sin\theta \, d\theta \tag{11.2}$$

$$V_{av} = \frac{1}{2\pi} \int_0^\pi V_m \sin\theta \, d\theta \tag{11.3}$$

$$V_{av} = \frac{1}{2\pi} \left[ -V_m \cos\theta \right]_0^\pi \tag{11.4}$$

$$V_{av} = V_m / \pi \tag{11.5}$$

For the 120-V supply, assuming ideal diodes

$$V_{av} = 120\sqrt{2}\,/\,\pi \qquad\qquad (11.6)$$

$$V_{av} = 54.019\text{V} \qquad\qquad (11.7)$$

only one diode conducts; therefore, the peak voltage and average voltage are reduced by 1.1V

$$V_{av} = 52.919\text{V} \qquad\qquad (11.8)$$

The full-wave rectifier has twice the average voltage as the waveform consists of positive half sine waves repeating every 180° or $\pi$ radians. For ideal diodes

$$V_{av} = 2V_m/\pi \qquad\qquad (11.9)$$

$$V_{av} = 108.04\text{V} \qquad\qquad (11.10)$$

For diodes with a forward voltage of 1.1V

$$V_{av} = 103.64\text{V} \qquad\qquad (11.11)$$

## Example 11.2

Calculate the RMS voltage for a half-wave rectifier and a full-wave rectifier that has a single-phase supply voltage of 120V. The anode-cathode voltage of each diode is 1.1V in forward conduction.

### Solution

For a half-wave rectifier

$$V_{RMS} = \sqrt{\frac{V_m^2}{2\pi}\int_0^\pi \sin^2\theta\; d\theta} \qquad\qquad (11.12)$$

$$V_{RMS} = \sqrt{\frac{V_m^2}{2\pi}\int_0^\pi \frac{1-\cos 2\theta}{2}\; d\theta} \qquad\qquad (11.13)$$

$$V_{RMS} = \sqrt{\frac{V_m^2}{2\pi}\left[\frac{\theta - \sin 2\theta\big/2}{2}\right]_0^\pi} \qquad\qquad (11.14)$$

$$V_{RMS} = \sqrt{\frac{V_m^2}{2\pi}\left[\pi/2\right]} \tag{11.15}$$

$$V_{RMS} = \sqrt{\frac{V_m^2}{4}} \tag{11.16}$$

$$V_{RMS} = \frac{V_m}{2} \tag{11.17}$$

For an ideal half-wave rectifier

$$V_{RMS} = 84.853 \text{V} \tag{11.18}$$

Taking into account the diode voltage of 1.1V

$$V_{RMS} = 58.9 \text{V} \tag{11.19}$$

A full-wave rectifier has an RMS voltage given by

$$V_{RMS} = \sqrt{\frac{V_m^2}{\pi}\int_0^\pi \sin^2\theta \, d\theta} \tag{11.20}$$

$$V_{RMS} = \sqrt{\frac{V_m^2}{\pi}\left[\frac{\theta - \sin 2\theta/2}{2}\right]_0^\pi} \tag{11.21}$$

$$V_{RMS} = \sqrt{\frac{V_m^2}{\pi}\left[\pi/2\right]} \tag{11.22}$$

$$V_{RMS} = \frac{V_m}{\sqrt{2}} \tag{11.23}$$

For an ideal full-wave rectifier

$$V_{RMS} = 84.853 \text{V} \tag{11.24}$$

For a half-wave with one conducting diode

$$V_{RMS} = 84.303 \text{V} \tag{11.25}$$

For a full wave with two conducting diodes

$$V_{RMS} = 83.753\text{V} \tag{11.26}$$

Note that the form factor for the single-phase and half-wave rectifier is

$$F_f = \frac{V_m}{2} \bigg/ \frac{V_m}{\pi} = \frac{\pi}{2} = 1.5708 \tag{11.27}$$

For the full-wave and single-phase supply, the form factor is

$$F_f = \frac{V_m}{\sqrt{2}} \bigg/ \frac{2V_m}{\pi} = \frac{\pi}{2\sqrt{2}} = 1.1107 \tag{11.28}$$

## Example 11.3

Calculate the average and RMS diode currents in a full-wave rectifier that is supplied from a single-phase voltage. The output current is smoothed by inductance in the load and is 1,000A.

### Solution

As each diode conducts for half the cycle, the average and RMS currents are

$$V_{av} = I_d/2 = 500\text{A} \tag{11.29}$$

## Example 11.4

A permanent magnet dc motor is powered from a 160-V dc power supply and rotates at 1,500 rpm with an armature current of 100A. It has an armature resistance of 0.096Ω. A dc chopper circuit supplies power to the motor with at a duty cycle of 0.75.
    Calculate:

(a) The back emf constant of the motor expressed in units of V radians$^{-1}$;
(b) The circuit power loss and efficiency.

### Solution

The equation governing the characteristic behavior of a dc motor is

$$V_a = I_a R_a + k_m \omega \tag{11.30}$$

Under steady-state conditions, the voltage across the armature inductance is zero. The armature voltage has two terms on the right side of the equation that are the voltage across the armature resistance and the back emf voltage.

The dc-to-dc converter (chopper) is controlled by the duty cycle, $\delta$, which has a range from 0 to 1.

$$V_a = \delta V_d \tag{11.31}$$

$$V_a = 160 \times 0.75 = 120 \,\text{V} \tag{11.32}$$

The voltage across the armature resistance is

$$V_{Ra} = 100 \times 0.096 = 9.6 \,\text{V} \tag{11.33}$$

The motor back emf is

$$V_{emf} = 120 - 9.6 = 110.4 \,\text{V} \tag{11.34}$$

The back emf constant in units of V/rpm is

$$k_m = \frac{V_{emf}}{f} \tag{11.35}$$

$$k_m = \frac{110.4}{1500} = 6.7115 \,\text{V} \, rpm^{-1} \tag{11.36}$$

Expressed in the correct units of V/radian, the constant is

$$k_m = \frac{6.7115 \times 2 \times \pi}{60} = 0.70283 \,\text{V} \, rad^{-1} \tag{11.37}$$

The power loss in the armature resistance is

$$P = I_a^2 R_a \tag{11.38}$$

$$P = 100^2 \times 0.096 = 960 \,\text{W} \tag{11.39}$$

Assuming that this is the major power loss (there are losses in the power semiconductors and there are the iron, friction, and windage losses of the motor), the efficiency is

$$\eta = \frac{V_a I_a - I_a^2 R_a}{V_a I_a} \tag{11.40}$$

$$\eta = \left[ \frac{120 \times 100 - 960}{120 \times 100} \right] \times 100 = 92\% \tag{11.41}$$

## Example 11.5

Calculate the voltages V1 and V2 for the transmission circuit shown in Figure 11.1, the firing angle of the rectifier, and the firing advance angle of the inverter. The system specification is given in Table 11.1.

The resistance of the cables, $R$, is 0.002Ω. Ignoring any power loss in the two converters, calculate the system efficiency.

## Solution

Each end of this dc-to-dc transmission system has six phase-controlled thyristors. The maximum mean rectified voltage with a zero delay angle and no load current ($I_d$ is zero) is

$$V_d = \frac{pV_m}{\pi} \sin\frac{\pi}{p} \cos 0 \qquad (11.42)$$

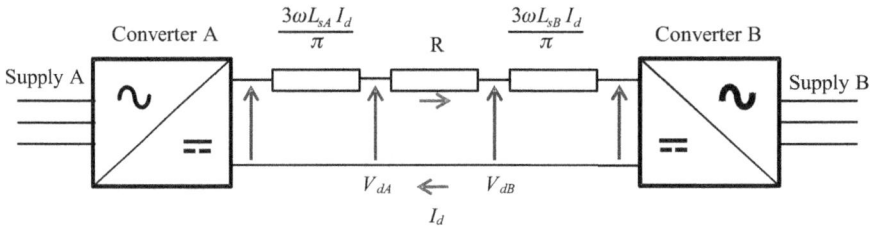

**Figure 11.1**    Diagram of an ac-to-ac power transmission via a dc link.

**Table 11.1**
System Specification

|                              | Converter 1                | Converter 2                |
|------------------------------|----------------------------|----------------------------|
| Type                         | Six-pulse fully controlled | Six-pulse fully controlled |
| Supply voltage               | 208V                       | 415V                       |
| Frequency                    | 60 Hz                      | 50 Hz                      |
| Source impedance per phase   | 0.011111 mH                | 0.011666 mH                |
| Mode                         | Rectifier                  | Inverter                   |
| Power                        |                            | 1.6 MW                     |
| Direct current               |                            | 10,000A                    |

For the converter A

$$V_d = \frac{6 \times 208 \times \sqrt{2}}{\pi} \sin\frac{\pi}{6} \cos 0 = 280.9\text{V} \tag{11.43}$$

For the converter B

$$V_d = \frac{6 \times 415 \times \sqrt{2}}{\pi} \sin\frac{\pi}{6} \cos 0 = 560.45\text{V} \tag{11.44}$$

From the power that is delivered to converter B, the input voltage is

$$V_d = P/I_d = 1.6\times10^6/10^4 = 160.00\text{V} \tag{11.45}$$

The voltage across the resistance of the cables is

$$V = IR = 10^4 \times 0.002 = 20.0\,\text{V} \tag{11.46}$$

The reactance of the converter A is

$$X_A = p\omega L_{sA}/2\pi = 6 \times 2\pi \times 60 \times 0.011111 \times 10^{-3}/2\pi = 0.004\Omega \tag{11.47}$$

The reactance of the converter B is

$$X_B = p\omega L_{sB}/2\pi = 6 \times 2\pi \times 50 \times 0.011666 \times 10^{-3}/2\pi = 0.0035\Omega \tag{11.48}$$

The voltage across the reactance at the converter A is

$$V = I_d X_A = 0.004 \times 10^4 = 40\text{V} \tag{11.49}$$

The voltage across the reactance at the converter B is

$$V = I_d X_B = 0.0035 \times 10^4 = 35\text{V} \tag{11.50}$$

The voltages of the converters are

$$V_{dB} = 160 + 35 = 195\text{V} \tag{11.51}$$

$$V_{dA} = 195 + 40 = 235\text{V} \tag{11.52}$$

The rectifier delay angle is

$$\alpha = \cos 235/280.9^{-1} = 0.57976\,rad\,(33.218^0) \tag{11.53}$$

The inverter extinction angle is

$$\beta = \cos 160 / 560.45^{-1} = 1.2813 \, rad \, (73.412^0) \qquad (11.54)$$

or the delay angle

$$\alpha = 1.8603 \, rad \, (106.59^0) \qquad (11.55)$$

The efficiency assuming no power loss in the converters is

$$\eta = \frac{100(V_{dA}I_d - 1_d^2 R)}{V_{dA}I_d} = \frac{100(160 \times 10^4 - 10^{4^2} 0.002)}{160 \times 10^4} \qquad (11.56)$$

$$\eta = 87.5\% \qquad (11.57)$$

## Example 11.6

The supply to a single phase to single-phase cycloconverter is 220V and 60 Hz. During an output half-cycle, the delay angles are 100°, 65°, 46°, 35°, 46°, 65°, and 100°. Calculate the average voltage over the positive half-cycle of the output voltage for a frequency of 8.5714 Hz. Also calculate the average power dissipated in a resistive load of 2.0953Ω.

The average voltages for each half-cycle are

$$V_{av} = \frac{V_m}{\pi}(1 + \cos \alpha) \qquad (11.58)$$

$$V_{av} = \frac{220\sqrt{2}}{\pi}(1 + \cos 100) = 81.838V \qquad (11.59)$$

$$V_{av} = \frac{220\sqrt{2}}{\pi}(1 + \cos 65) = 140.89V \qquad (11.60)$$

$$V_{av}\frac{220\sqrt{2}}{\pi}(1 + \cos 46) = 167.83V \qquad (11.61)$$

$$V_{av} = \frac{220\sqrt{2}}{\pi}(1 + \cos 35) = 180.16 \, V \qquad (11.62)$$

The average voltage over a half-cycle is

$$V_{av} = \frac{(81.838 + 140.89 + 167.83 + 180.16 + 167.83 + 140.89 + 81.838)}{7} \qquad (11.63)$$

$$V_{av} = 137.32 \text{ V} \tag{11.64}$$

The average power is

$$P_{av} = 137.32^2 \times 2.0593 = 39.513 \text{ kW} \tag{11.65}$$

## Example 11.7

For a dc-to-dc and step-down converter, calculate the average output (load) current if the supply voltage is 160V, the output power is 1,200W, and the duty cycle is 0.3.

Calculate the output current at the boundary of continuous and discontinuous current for a switching frequency of 25 kHz. The inductor has a value of 50 μH. Also calculate the peak current.

Calculate the peak inductor current and the time during the off-period when the inductor current is zero for a duty cycle of 0.02.

### Solution

The output voltage is

$$V_o = \delta \, V_d = 0.3 \times 160 = 48 \text{ V} \tag{11.66}$$

Assuming no losses in the power converter, then the input power and output power are equal.

$$V_d \, i_d = V_o \, i_o \tag{11.67}$$

The output current is

$$i_o = V_d i_d / V_o = 1200/48 = 25\text{A} \tag{11.68}$$

The period of operation is

$$T_p = 1/f = 1/25000 = 40 \times 10^{-6} s \tag{11.69}$$

At the boundary between continuous and discontinuous current

$$I_{LB} = I_{OB} = \delta T_p (V_d - V_o)/2L \tag{11.70}$$

$$I_{OB} = 0.3 \times 40 \times 10^{-6} (160 - 48)/2 \times 50 \times 10^{-6} = 13.44\text{A} \tag{11.71}$$

The peak current is simply twice the average current

$$I_{OB} = 26.88\text{A} \tag{11.72}$$

Under discontinuous operation, the current rises to a peak and then falls to zero.

The proportion of the time, $\gamma$, that the current takes to fall from is given by

$$\frac{V_o}{V_d} = \frac{\delta}{(\gamma + \delta)} \tag{11.73}$$

$$\gamma = \frac{\delta(V_d - V_o)}{V_d} = \frac{0.02(160 - 48)}{160} = 0.046667 \tag{11.74}$$

The proportion of the cycle left when the current is zero is

$$(1 - \delta - \gamma)T_p = (1 - 0.02 - 0.046667)\,40 \times 10^{-6}\,s = 37.333\ \mu s \tag{11.75}$$

## Example 11.8

A step-up converter has a maximum input power of 1.5 kW at a voltage of 24V. The output voltage is 400V when the switching frequency is 50 kHz.

Calculate the transistor on time, the peak current, and the value of the inductor.

### Solution

The duty cycle, $\delta$, is

$$\delta = 1 - (V_d/V_o) = 1 - (24/400) = 0.94 \tag{11.76}$$

The period of a cycle is

$$T_p = 1/50\,000 = 20\ \mu s \tag{11.77}$$

The transistor on time is

$$t_1 = \delta T_p = 0.94 \times 20 = 18.8\ \mu s \tag{11.78}$$

The input current is

$$i_d = P/V_d = 1500/24 = 62.4 \text{A} \tag{11.79}$$

The output current is

$$i_o = P/V_o = 1500/400 = 3.75 \text{A} \tag{11.80}$$

The inductance is

$$L = V_d T_p \, \delta^2 \left( \frac{V_o}{V_o - V_d} \right) \Big/ 2 i_o \tag{11.81}$$

$$L = 24 \times 20 \times 10^{-6} \times 0.94^2 \left( \frac{400}{400 - 24} \right) \Big/ 2 \times 3.75 \tag{11.82}$$

$$L = 60.15 \, \mu H \tag{11.83}$$

## Example 11.9

The gate characteristic of a power MOSFET is given by the following equation, where the gate-source voltage is $v_{GS}$ and the gate charge is $q_G$.

$$v_{GS} = k_1 + k_2 q_G \tag{11.84}$$

Draw a plot of this characteristic with $v_{GS}$ on the vertical axis and $q_G$ on the horizontal axis using the parameter values shown in Table 11.2.

Calculate the average power needed to switch on the transistor if it is turned on and off at a frequency of 50 kHz. The voltage applied to the gate is given by the following equations.

$$v_{GS} = 0 \text{ V} \quad \text{for } t = 0 \tag{11.85}$$

$$v_{GS} = 10 \text{ V} \quad \text{for } t > 0 \tag{11.86}$$

The electronic drive circuit to the gate has an output resistance of $0.1\Omega$. Calculate the time taken for the MOSFET gate voltage to reach 8V.

### Solution

From the stated values of gate charge, the gate voltages are calculated (Table 11.3).

### Table 11.2
Parameter Values for $k_1$ and $k_2$

| $k_1$ (V) | $k_2$ (V nC$^{-1}$) | Range of $q_G$ (nC) |
|---|---|---|
| 0 | 2.5 | $0 \le q_G \le 2$ |
| 4.9967 | 0.0016667 | $2 \le q_G \le 8$ |
| −4.9699 | 1.2475 | $8 \le q_G \le 12$ |

**Table 11.3**
Calculated Gate Charge

| $k_1$ (V) | $k_2$ (V nC$^{-1}$) | $q_G$ (nC) | $v_{GS}$ (V) |
|---|---|---|---|
| 0 | 2.5 | 0 | 0 |
| 0 | 2.5 | 2 | 5 |
| 4.9967 | 0.001667 | 2 | 5 |
| 4.9967 | 0.001667 | 8 | 5.01 |
| −4.9699 | 1.2475 | 8 | 5.01 |
| −4.9699 | 1.2475 | 12 | 10 |

**Table 11.4**
Calculated Times to Charge the Gate to 8V

| First coordinate | | Last coordinate | | | | |
|---|---|---|---|---|---|---|
| $q_G$ (nC) | $v_{GS}$ (V) | $q_G$ (nC) | $v_{GS}$ (V) | $C = q_G/V$ (nC) | $V_o$ (V) | $t$ (ns) |
| 0 | 0 | 2 | 5 | 0.4 | 10 | 0.027726 |
| 2 | 5 | 8 | 5.01 | 600 | 5 | 0.12012 |
| 8 | 5.01 | 12 | 10 | 0.80161 | 4.99 | 0.07329 |

Table 11.4 shows the start and end coordinates for each of the three straight lines using (11.90). From the values of $q_G$ the plot is made (Figure 11.2).

The charge and hence energy stored in the gate occurs very quickly compared to the switching frequency of 50 kHz. At 10V the gate has 12 nC of stored charge and the energy is

$$E = q_G \, v_{GS} = 12 \times 10^{-9} \times 10 = 120 \; nJ \tag{11.87}$$

$$P_{av} = E/T_p = E \, f = 120 \times 10^{-9} \times 50 \times 10^3 = 6 \; mW \tag{11.88}$$

As this energy is stored, it could be recovered but is not in practice.

The gate-drive circuit dissipates power, particularly in the output stage, every time the transistor is turned on and off. The output stage is typically a MOSFET that presents a series resistance to the current charging the gate capacitance. A series resistance and capacitance circuit is formed where the current decays exponentially with time. The equation governing the voltage rise is

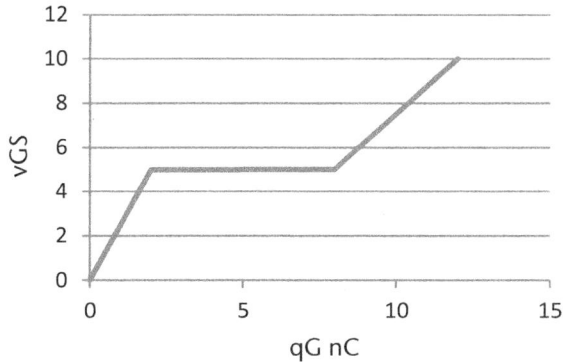

**Figure 11.2** Gate-charge characteristic.

$$v_{GS} = V_o(1 - \exp(t/RC)) \tag{11.89}$$

where $V_o$ is the difference between the final voltage (10V) and the initial voltage on the capacitor.

Rearranging this equation,

$$t = RC\ln(V_o/V_o - v_{GS})) \tag{11.90}$$

The total time taken to charge the gate to 8V is 0.22114 ns. Of the three parts in the gate characteristic, it is the one with the lowest slope that forms the largest capacitance. While charging this capacitance, the change in gate voltage is small.

## Example 11.10

An inductor, $L$, is connected in parallel with the drain and source of an n-channel power MOSFET that is turned off. The drain to source voltage, $v_{DS}$, is negative. There is a current, $i(t)$, flowing through the inductor. Derive a second-order differential equation for the time, $t$, behavior of the current, $i$. Define all the symbols used in your equations. By making a linear approximation for the relationship between current and voltage, show that the voltage decays exponentially with time.

### Solution

The fundamental equations for the inductor and diode are given next.

The equation for the inductor is

$$v_{DS} = L\frac{di_D}{dt} \tag{11.91}$$

As the MOSFET has a negative drain-to-source voltage, the parasitic diode is forward biased and in parallel.

$$i_D = I_o \left(\exp(-q\, v_{DS}/kT) - 1\right) \qquad (11.92)$$

Differentiating this equation,

$$\frac{di_D}{dt} = \left[-I_o q/kT \exp(-q v_{DS}/kT)\right] \frac{dv_{DS}}{dt} \qquad (11.93)$$

Rearranging this equation and substituting the exponential function,

$$\exp(-q v_{DS}/kT) = (i_D + i_o)/I_o \qquad (11.94)$$

$$\frac{dv_{DS}}{dt} = -\left[kT/q(i_D + I_o)\right] \frac{di_D}{dt} \qquad (11.95)$$

Differentiating the inductor equation twice and eliminating the first derivative of $v_{DS}$,

$$\frac{dv_{DS}}{dt} = L \frac{d^2 i_D}{dt^2} \qquad (11.96)$$

$$L\frac{d^2 i_D}{dt^2} + \frac{kT}{q\,(i_D + I_o)} \frac{di_D}{dt} = 0 \qquad (11.97)$$

Expressing the exponential function as a series and ignoring terms with an order higher than 2, a linear approximation is formed.

$$i_D = I_o \left[(1 - q v_{DS}/kT + \cdots) - 1\right] \qquad (11.98)$$

$$i_D = -(I_o q/kT)\, v_{DS} \qquad (11.99)$$

This equation is differentiated with respect to time.

$$\frac{di_D}{dt} = -(I_o q/kT) \frac{dv_{DS}}{dt} \qquad (11.100)$$

Combining this equation with that for the inductance and eliminating the current derivative gives

$$v_{DS} = -(L I_o q/kT) \frac{dv_{DS}}{dt} \qquad (11.101)$$

Rearranging and integrating this equation gives

$$- (kT/LI_oq) \, dt = \frac{dv_{DS}}{v_{DS}} \tag{11.102}$$

$$- (kT/LI_o q) \, t = \ln(v_{DS}) \tag{11.103}$$

or

$$v_{DS} = V_o - V_o \exp(-(kT/Li_oq) \, t) \tag{11.104}$$

where $V_o$ is the initial voltage.

## Example 11.11

Determine the Fourier series expansion for the half-wave rectified voltage, $v$, shown in Figure 11.3. Express the series in terms of the harmonic number, $r$.

Plot three waveforms that are the half-wave rectified waveform, the Fourier series that includes the first term and the Fourier series that includes all terms up to the twentieth. Also plot waveforms for the differences between the rectified values and those for the waveforms that include all Fourier components up to the second, fourth, sixth, eighth, and tenth terms. Plot the maximum values, absolute of the minimum values and average values showing the effects of adding

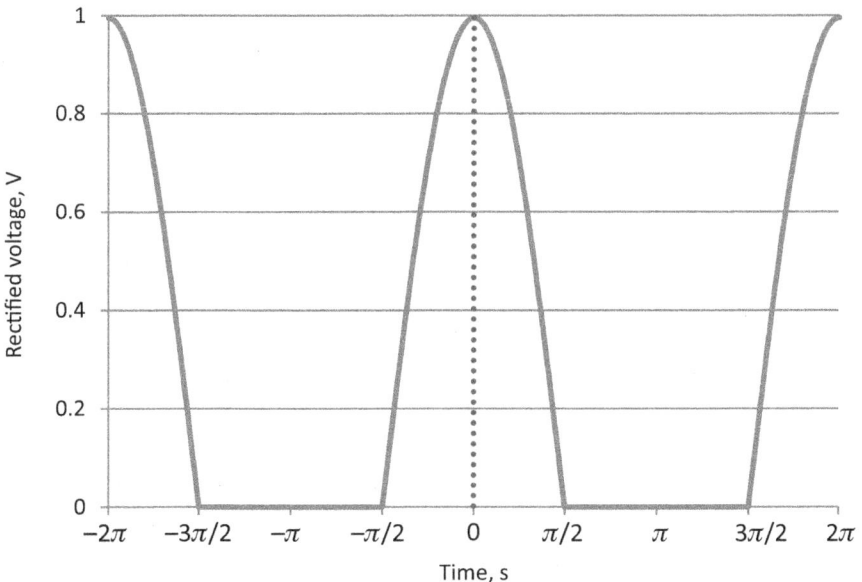

**Figure 11.3** Rectified voltage, $v$ (cosine).

Fourier components up to the twentieth. Determine an expression for the ratio, $F$, of the fundamental to the harmonic number, $r$. Show that this expression approximates to a power relationship.

## Solution

The function of the waveform is a cosine and therefore it is an even function centered about the time $t = 0$. The coefficients, $b_r$, are all zero and only the amplitudes, $a_r$, are calculated as they are cosines.

$$f(t) = 0 \quad for -\pi < t \leq \pi/2 \tag{11.105}$$

$$f(t) = \cos t \quad for -\pi/2 < t < \pi/2 \tag{11.106}$$

$$f(t) = 0 \quad for \pi/2 \leq t \leq \pi \tag{11.107}$$

The dc term, $a_0$, is

$$a_0 = \frac{1}{\pi} \int_{-\pi}^{\pi} f(t)\, dt \tag{11.108}$$

The function between $-\pi$ and $-\pi/2$ and between $\pi$ and $\pi/2$ is zero; hence,

$$a_o = \frac{2}{\pi} \int_o^{\pi} f(t)\, dt = \frac{2}{\pi} \int_o^{\pi/2} |\cos t|\, dt \tag{11.109}$$

$$a_0 = \frac{2}{\pi} \left[\sin t\right]_0^{\pi/2} = \frac{2}{\pi} \tag{11.110}$$

For the components, $a_r$, the coefficient equation is

$$a_r = \frac{1}{\pi} \int_{-\pi}^{\pi} f(x) \cos rx\, dx, (r = 1, 2, 3, ...) \tag{11.111}$$

As a cosine is an even function, then the integral can be simplified to

$$a_r = \frac{2}{\pi} \int_0^{\pi} f(x) \cos rx\, dx, (r = 1, 2, 3, ...) \tag{11.112}$$

Substituting in the cosine

$$a_r = \frac{2}{\pi} \int_0^{\pi} |\cos t| \cos rt\, dt \tag{11.113}$$

Because the half-rectified waveform exists only from $t = 0$ to $t = \pi/2$,

$$a_r = \frac{2}{\pi} \int_0^{\pi/2} \cos t \cos rt\, dt \tag{11.114}$$

Using the identity relating the cosines to their sum and difference, then

$$a_r = \frac{2}{\pi}\int_0^{\pi/2}\left[\frac{\cos(r+1)t}{2} + \frac{\cos(r-1)t}{2}\right]dt \qquad (11.115)$$

Integrating this equation

$$a_r = \frac{1}{\pi}\left[\frac{1}{(r+1)}\sin(r+1)t + \frac{1}{(r-1)}\sin(r-1)t\right]_0^{\pi/2} \qquad (11.116)$$

The sinusoidal terms are zero for the lower limit of $t = 0$. Substituting in the upper limit, the first sine term is

$$\sin(rt+1) = \sin rt \cos t + \cos rt \sin t \qquad (11.117)$$

and the second term is

$$\sin(rt-1) = \sin rt \cos t - \cos rt \sin t \qquad (11.118)$$

Eliminating terms

$$a_r = \frac{1}{\pi}\left[\frac{(r-1)(\sin rt\cos t + \cos rt \sin t) + (r+1)(\sin rt\cos t - \cos rt \sin t)}{(r+1)(r-1)}\right]_0^{\pi/2} \qquad (11.119)$$

As $\cos(\pi/2)$ is zero, the first term is zero leaving.

$$a_r = \frac{1}{\pi}\left[\frac{2\,(r\sin rt\cos t - \cos rt\sin t)}{(r^2-1)}\right]_0^{\pi/2} \qquad (11.120)$$

$$a_r = \frac{-2}{\pi(r^2-1)}\left(\cos\frac{r\pi}{2}\sin\frac{\pi}{2}\right) \qquad (11.121)$$

$$a_r = \frac{-2}{\pi(r^2-1)}\cos\frac{r\pi}{2} \qquad (11.122)$$

Ignore the first term for the moment as it has to be determined separately. For $r = 2$, $\cos(2\pi/2) = -1$, for $r = 3$, $\cos(3\pi/2) = 0$, for $r = 4$, $\cos(4\pi/2) = 1$ ... The coefficients alternate between negative and positive values. Hence,

$$a_r = \frac{-2}{\pi(r^2-1)}\,(-1)^{r/2} \quad r = 2,4,6,... \qquad (11.123)$$

The first term is determined as follows.

$$a_1 = \frac{2}{\pi} \int_0^{\pi/2} \cos t \cos t \, dt \tag{11.124}$$

Using the identity for the square of a cosine

$$a_1 = \frac{2}{\pi} \int_0^{\pi/2} \frac{1 + \cos 2t}{2} \, dt \tag{11.125}$$

$$a_1 = \frac{1}{\pi} \left[ t + \frac{1}{2} \sin 2t \right]_0^{\pi/2} \tag{11.126}$$

$$a_1 = \frac{1}{\pi} \left[ \frac{\pi}{2} \right] = \frac{1}{2} \tag{11.127}$$

Therefore, the Fourier series of a half-wave rectified cosine is

$$v(t) = \frac{1}{\pi} + \frac{1}{2} \cos t + \sum_{r=2}^{\infty} \frac{-2}{\pi(r^2 - 1)} (-1)^{r/2} \cos rt \tag{11.128}$$

$$\text{for } r = 2, 4, 6, \ldots \tag{11.129}$$

The Fourier series for $r = 2$ is plotted in Figure 11.4. The waveform with terms up to $r = 20$ is also shown in the plot together with the half rectified

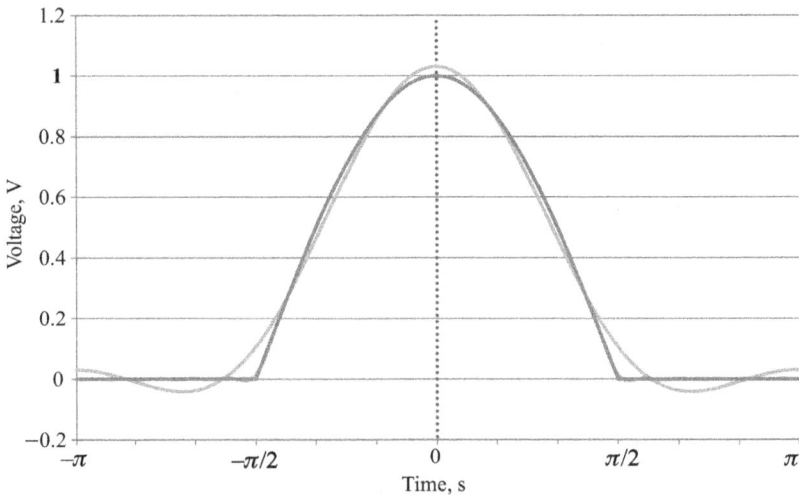

**Figure 11.4**  Waveforms for Fourier series of the first term and up to the twentieth term. The rectified waveform is included for comparison with the values from the Fourier series.

waveform. The sign of the terms shown in the summation of the Fourier series alternates with positive values for $r = 2, 6, 10, \ldots$ and negative values for $r = 4, 8, 12, \ldots$. Notice that the waveform with the higher harmonics ($r = 20$) is not easily distinguishable from the rectified waveform. Some differences can be seen around the times $\pi/2$ and $-\pi/2$.

The differences between the rectified waveform and the Fourier series are shown in Figure 11.5. The difference is highest at the times of $\pi/2$ and $-\pi/2$.

The magnitude ratio, $F$, of the fundamental to the harmonic number, $r$, is given by

$$F = \frac{|a_r|}{|a_1|} = \frac{2/\pi(r^2 - 1)}{1/2} = \frac{4}{\pi(r^2 - 1)} \tag{11.130}$$

Fitting a power relationship to the data shown in Figure 11.6 (dotted curve),

$$F = 1.6608 \, r^{-2.1012} \tag{11.131}$$

The analytically determined ratio is compared with that calculated from the power expression in Figure 11.7. With only one term ($r = 2$), there is nearly a 4% difference between the two values (Figure 11.8). From the tenth term, there is very little difference between the values as expected from the presence of the inverse ($r^2 - 1$) factor in the analytical equation.

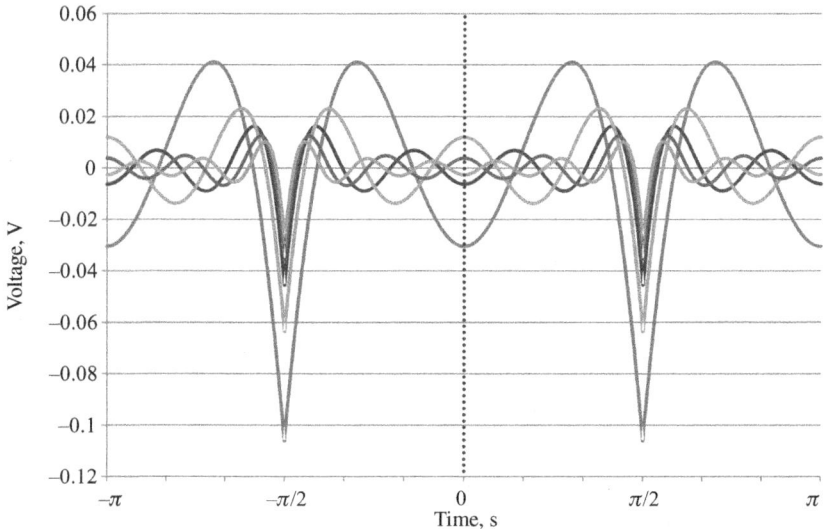

**Figure 11.5** Differences between the rectified waveform and the Fourier series that includes terms up to the second, fourth, sixth, eighth, and tenth harmonics.

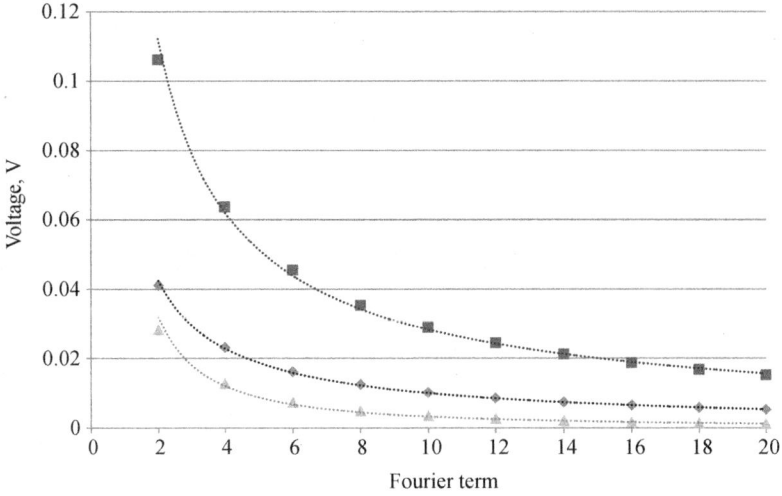

**Figure 11.6** Shown in the upper curve are the absolute minimum differences as a function of the number of terms included in the Fourier series that occur at $\pi/2$ and $-\pi/2$. In the middle curve are plotted the maximum differences and in the lower curve are the average differences.

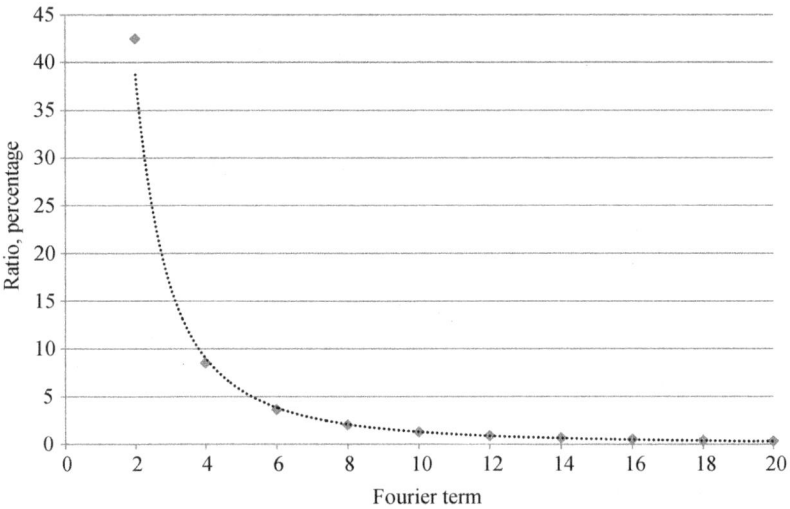

**Figure 11.7** Ratio of the magnitude of the harmonic component, $r$, to the fundamental component expressed as a percentage.

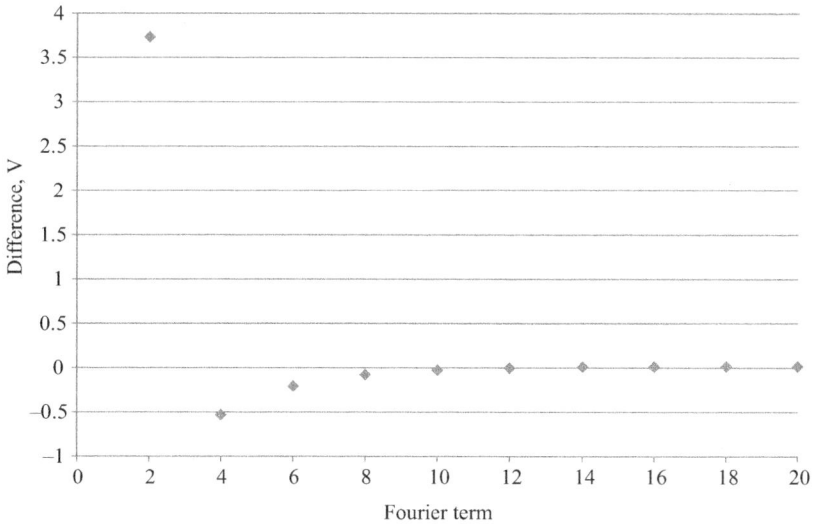

**Figure 11.8**   Difference between the ratio of the magnitude of the harmonic component, *r*, to the fundamental component from the analytical and power relationships expressed as a percentage.

## Example 11.12

Determine the Fourier series expansion for the full-wave rectified voltage, *v*, shown in Figure 11.9. Express the series in terms of the harmonic number, *r*.

Plot three waveforms that are the full-wave rectified waveform, the Fourier series that includes the first term, and the Fourier series that includes all harmonics up to the twentieth. Also plot waveforms for the differences between the rectified values and those for the waveforms that include all Fourier components up to the second, fourth, sixth, eighth, and tenth terms. Plot the maximum values, absolute of the minimum values, and average values showing the effects of adding Fourier components up to the twentieth. Determine an expression for the ratio, *F*, of the fundamental to the harmonic number, *r*. Show that this expression approximates to a power relationship.

### Solution

The rectified waveform is a cosinusoidal function and therefore is an even function centered about the time *t* = 0.

$$f(t) = -\cos t \quad \text{for} -\pi < t \le \pi/2 \tag{11.132}$$

$$f(t) = \cos t \quad \text{for} -\pi/2 < t < \pi/2 \tag{11.133}$$

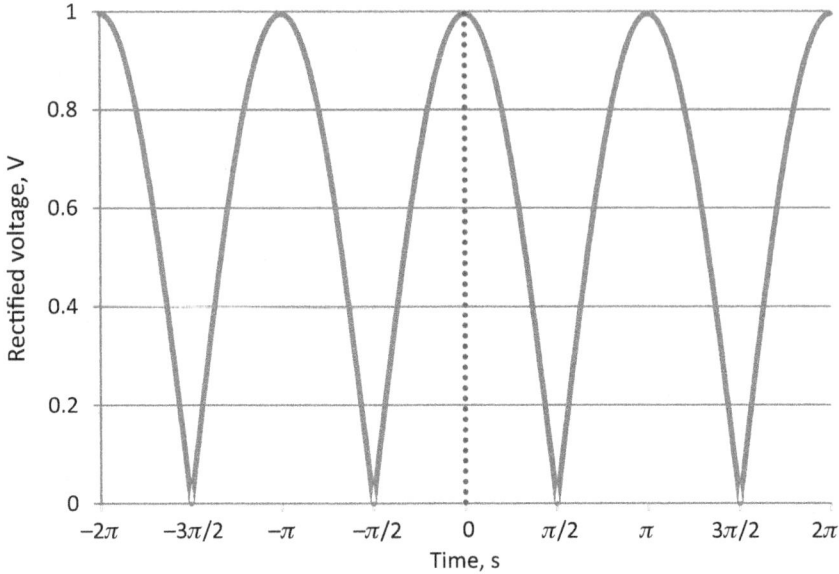

**Figure 11.9**   Full rectified voltage, $v$ (cosine).

$$f(t) = -\cos t \quad or \, \pi/2 \le t < \pi \tag{11.134}$$

The coefficients, $b_r$, are all zero and only the cosine amplitudes, $a_r$, are calculated.

The dc term, $a_0$, is

$$a_0 = \frac{1}{\pi} \int_{-\pi}^{\pi} f(t) \, dt \tag{11.135}$$

For the rectified waveform that is a cosine, the integral is

$$\int_{0}^{\pi} |\cos t| \, dt = \int_{-\pi}^{0} |\cos t| \, dt \tag{11.136}$$

Hence,

$$a_0 = \frac{2}{\pi} \int_{0}^{\pi} |\cos t| \, dt \tag{11.137}$$

For the range of $t$ from $\pi/2$ to $\pi$ the cosine function is inverted and the integral is determined in two parts.

$$a_0 = \frac{2}{\pi}\left[\int_{\pi/2}^{\pi} -\cos t \; dt + \int_0^{\pi/2} \cos t \; dt\right] \qquad (11.138)$$

$$a_0 = \frac{2}{\pi}\left\{[-\sin t]_{\pi/2}^{\pi} + [\sin t]_0^{\pi/2}\right\} \qquad (11.139)$$

$$a_0 = \frac{2}{\pi}\{-\sin \pi + \sin \pi/2 + \sin \pi/2 - \sin 0\} \qquad (11.140)$$

$$a_0 = \frac{4}{\pi} \qquad (11.141)$$

For the harmonic components, $a_r$, the coefficient equation is

$$a_r = \frac{1}{\pi}\int_{-\pi}^{\pi} f(x)\cos rx \; dx, \; (r = 1,2,3,...) \qquad (11.142)$$

As a cosine is an even function, then the integral can be simplified to

$$a_r = \frac{2}{\pi}\int_0^{\pi} f(x)\cos rx \; dx, \; (r = 1,2,3,...) \qquad (11.143)$$

Substituting in the cosine function

$$a_r = \frac{2}{\pi}\int_0^{\pi} |\cos t|\cos rt \; dt \qquad (11.144)$$

Splitting the integral into two parts

$$a_r \frac{2}{\pi}\left[\int_{\pi/2}^{\pi} -\cos t \cos rt + \int_0^{\pi/2} \cos t \cos rt\right]dt \qquad (11.145)$$

$$a_r \frac{2}{\pi}\left[-\int_{\pi/2}^{\pi} \cos rt \cos t + \int_0^{\pi/2} \cos rt \cos t\right]dt \qquad (11.146)$$

Using a trigonometric identity

$$a_r \frac{2}{\pi}\left[-\int_{\pi/2}^{\pi} \frac{\cos(rt+t)\cos(rt-t)}{2} + \int_0^{\pi/2} \frac{\cos(rt+t)\cos(rt-t)}{2}\right]dt \qquad (11.147)$$

$$a_r = \frac{1}{\pi} \left\{ -\left[ \frac{1}{(r+1)} \sin(rt+t) + \frac{1}{(r-1)} \sin(rt+t) \right]_{\pi/2}^{\pi} \right.$$

$$\left. + \left[ \frac{1}{(r+1)} \sin(rt+t) + \frac{1}{(r-1)} \sin(rt-t) \right]_{0}^{\pi/2} \right\} \qquad (11.148)$$

$$a_r = \frac{1}{\pi} \left\{ -\left[ \frac{1}{(r+1)} \sin(r+1)\pi + \frac{1}{(r-1)} \sin(r-1)\pi - \frac{1}{(r+1)} \sin(r+1)\pi/2 \right. \right.$$

$$\left. -\frac{1}{(r-1)} \sin(r-1)\pi/2 \right] + \left[ \frac{1}{(r+1)} \sin(r+1)\pi/2 + \frac{1}{(r-1)} \sin(r-1)\pi/2 \right.$$

$$\left. \left. -\frac{1}{(r+1)} \sin(r+1)0 - \frac{1}{(r-1)} \sin(r-1)0 \right] \right\} \qquad (11.149)$$

For angles of $(r+1)\pi$, $(r-1)\pi$, and 0, the sine terms are all zero. Hence,

$$a_r = \frac{2}{\pi} \left\{ \frac{1}{(r+1)} \sin(r+1)\pi/2 + \frac{1}{(r-1)} \sin(r-1)\pi/2 \right\} \qquad (11.150)$$

Table 11.5 shows the steps in simplifying the first three terms ($r = 2, 4$, and 6) for $a_r$.

**Table 11.5**
Steps in Simplifying the First Three Terms for $a_r$

| $r$ | $\dfrac{1}{(r+1)} \sin(r+1)\pi/2 + \dfrac{1}{(r-1)} \sin(r-1)\pi/2$ |
|---|---|
| 2 | $-1/(r+1) + 1/(r-1)$ |
| 4 | $1/(r+1) - 1/(r-1)$ |
| 6 | $-1/(r+1) + 1/(r-1)$ |
| $R$ | $\dfrac{1}{(r+1)} \sin(r+1)\pi/2 + \dfrac{1}{(r-1)} \sin(r-1)\pi/2$ |
| 2 | $\dfrac{-(r-1)+(r+1)}{(r^2-1)} = \dfrac{+2}{(r^2-1)}$ |
| 4 | $\dfrac{(r-1)-(r+1)}{(r^2-1)} = \dfrac{-2}{(r^2-1)}$ |
| 6 | $\dfrac{-(r-1)+(r+1)}{(r^2-1)} = \dfrac{+2}{(r^2-1)}$ |

The first component ($r = 2$) is positive, while the second ($r = 4$) is negative. This pattern repeats for the higher-order components.

Hence,

$$a_r = \frac{-4}{\pi(r^2 - 1)} (-1)^{r/2} \quad r = 2, 4, 6, \dots \tag{11.151}$$

The first term, $a_1$, is determined as follows.

$$a_1 = \frac{2}{\pi} \int_0^\pi |\cos t| \cos t \, dt \tag{11.152}$$

$$a_1 = \frac{2}{\pi} \left[ \int_{\pi/2}^\pi -\cos t \cos t + \int_0^{\pi/2} \cos t \cos t \right] dt \tag{11.153}$$

Using the identity for the square of a cosine

$$a_1 = \frac{2}{\pi} \left\{ -\int_{\pi/2}^\pi \frac{1 + \cos 2t}{2} \, dt + \int_0^{\pi/2} \frac{1 + \cos 2t}{2} \, dt \right\} \tag{11.154}$$

$$a_1 = \frac{1}{\pi} \left\{ -\left[ t + \frac{1}{2} \sin 2t \right]_{\pi/2}^\pi + \left[ t + \frac{1}{2} \sin 2t \right]_0^{\pi/2} \right\} \tag{11.155}$$

$$a_1 = \frac{1}{\pi} \left\{ -\left[ \pi + \frac{1}{2} \sin 2\pi \right] - \left[ -\left( \frac{\pi}{2} + \frac{1}{2} \sin \frac{2\pi}{2} \right) \right] \right.$$
$$\left. + \left[ \frac{\pi}{2} + \frac{1}{2} \sin \frac{2\pi}{2} - \left( 0 + \frac{1}{2} \sin 2 \times 0 \right) \right] \right\} \tag{11.156}$$

$$a_1 = \frac{1}{\pi} \left[ -\pi - \frac{1}{2} \times 0 + \frac{\pi}{2} + \frac{1}{2} \sin 0 + \frac{\pi}{2} + \frac{1}{2} \sin 0 \right] \tag{11.157}$$

$$a_1 = 0 \tag{11.158}$$

Therefore, $v(t)$, for the full rectified waveform is given by

$$v(t) = \frac{2}{\pi} + \sum_{r=2}^\infty \frac{-4}{\pi(r^2 - 1)} (-1)^{r/2} \cos rt \tag{11.159}$$

$$\text{for } r = 2, 4, 6, \dots \tag{11.160}$$

Note that for $r = 1$, there is no component, so that the fundamental component is for $r = 2$.

The Fourier series for $r = 2$ is plotted in Figure 11.10. The waveform with terms up to $r = 20$ is also shown in the plot together with the full rectified waveform. The sign of the terms shown in the summation of the Fourier series alternates with positive values for $r = 2, 6, 10, \ldots$ and negative values for $r = 4, 8, 12, \ldots$. Notice that the waveform with the higher harmonics ($r = 20$) is not easily distinguishable from the rectified waveform. Some differences can be seen around the times $\pi/2$ and $-\pi/2$.

The differences between the rectified waveform and the Fourier series are shown in Figure 11.11, where the difference is highest at the times of $\pi/2$ and $-\pi/2$. A summary of the maximum, absolute minimum, and mean differences are shown in Figure 11.12.

The magnitude ratio, $F$, of the fundamental to the harmonic number, $r$, is given by

$$F = \frac{|a_r|}{a_1} = \frac{4(-1)^{r/2}\big/\pi(r^2-1)}{-4\big/\pi(2^2-1)} = \frac{3}{(r^2-1)} \tag{11.161}$$

Figure 11.13 shows the ratio plotted as a function of component number.

Fitting a power relationship to the data shown in Figure 11.13 (dotted curve),

$$F = 2.5189\, r^{-2.03403} \tag{11.162}$$

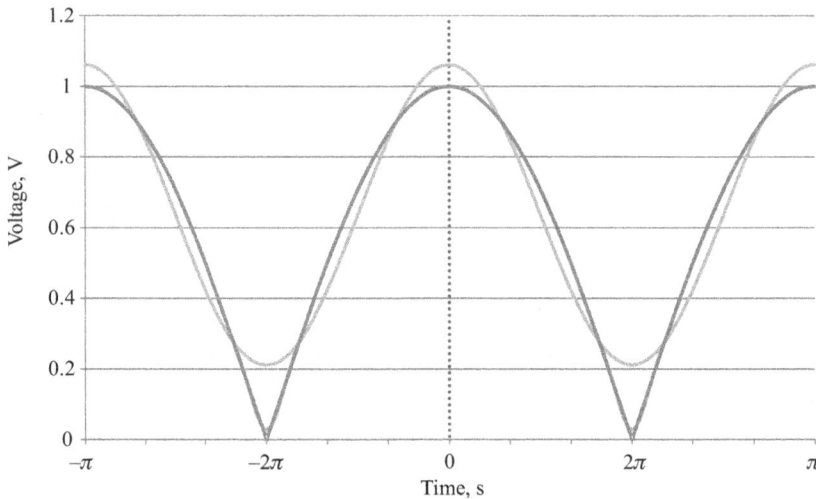

**Figure 11.10**    Fourier series of the first term ($r = 2$) and up to the twentieth term. The rectified waveform is included for comparison with the values from the Fourier series.

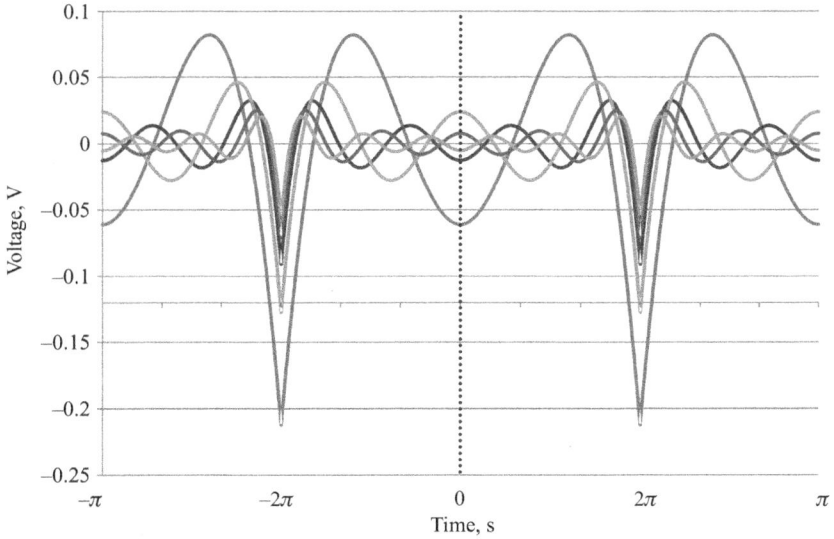

**Figure 11.11**    Differences between the rectified waveform and the Fourier series that includes terms up to the second, fourth, sixth, eighth, and tenth harmonics.

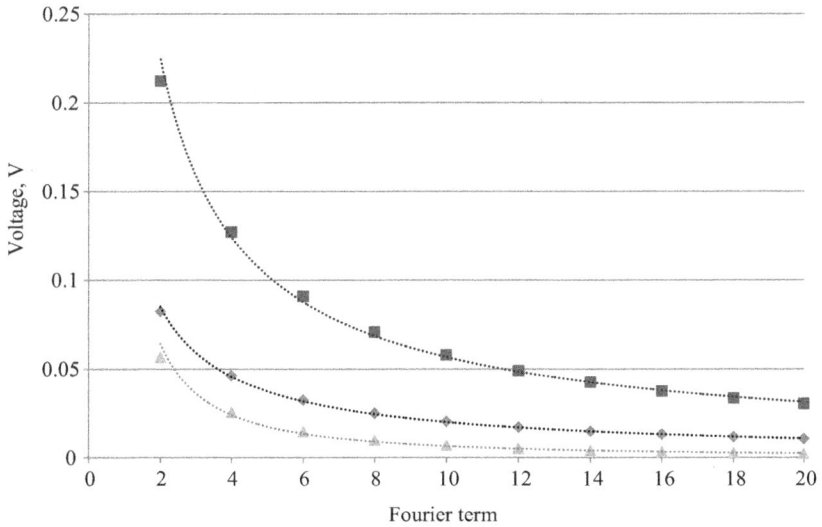

**Figure 11.12**    Shown in the upper curve are the absolute minimum differences as a function of the number of terms included in the Fourier series that occur at $\pi/2$ and $-\pi/2$. In the middle curve are plotted the maximum differences and in the lower curve are the average differences.

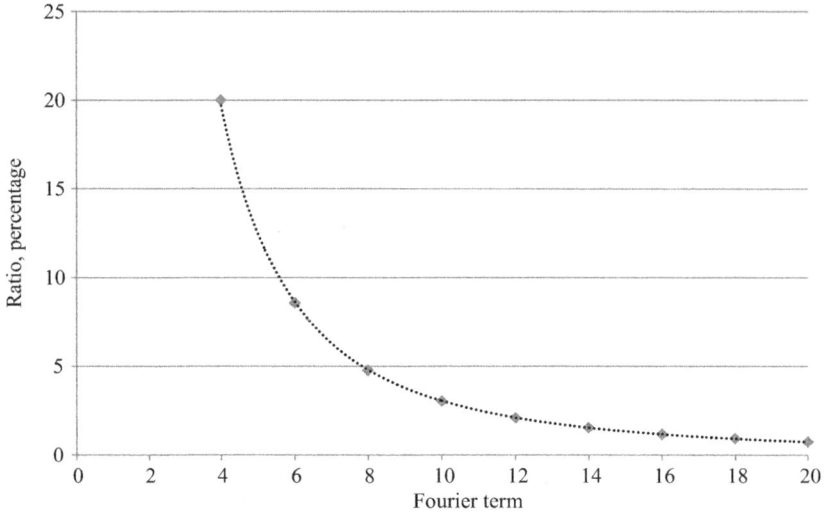

**Figure 11.13**   Ratio of the magnitude of the harmonic component, r, to the fundamental component (r = 2) expressed as a percentage.

The analytically determined ratio is compared with that calculated from the power expression in Figure 11.14. With only one term (r = 2), there is slightly more than a 0.3% difference between the two values. From the twelfth term, there is very little difference between the values as expected from the presence of the inverse $(r^2 - 1)$ factor in the analytical equation.

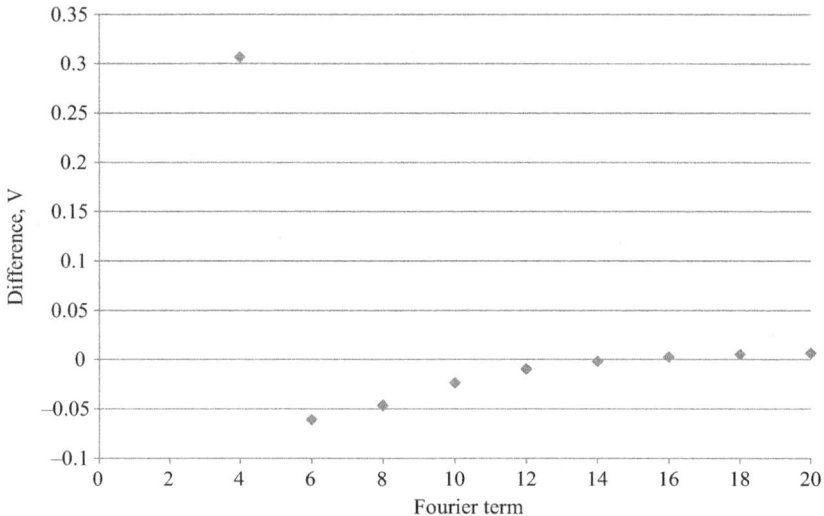

**Figure 11.14**   Difference between the ratio of the magnitude of the harmonic component, r, to the fundamental component (r = 2) from the analytical and power relationships expressed as a percentage.

## Example 11.13

Determine the Fourier series expansion for the half-wave rectified voltage, $v$, shown in Figure 11.15.

### Solution

This function is neither odd nor even. The analysis of this waveform proceeds in a similar manner to that of the cosine waveform.

$$f(t) = 0 \quad \text{for} -\pi \le t < 0 \tag{11.163}$$

$$f(t) = \sin t \quad \text{for } 0 < t < \pi \tag{11.164}$$

The dc term, $a_0$, is

$$a_0 = \frac{2}{\pi} \int_0^\pi f(t) \, dt = \frac{2}{\pi} \int_0^\pi \sin t \, dt \tag{11.165}$$

$$a_0 = \frac{2}{\pi} \left[ -\cos t \right]_0^\pi = \frac{2}{\pi} \tag{11.166}$$

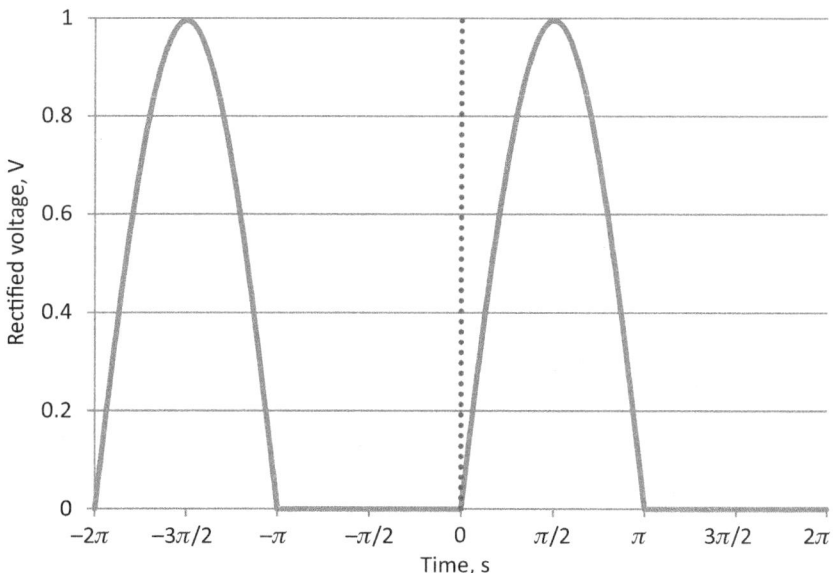

**Figure 11.15**  Rectified voltage, $v$ (sine).

As the waveform is zero from $t = -\pi/2$ to $0$, the Fourier coefficients are given by

$$a_r \frac{1}{\pi} \int_0^\pi \sin t \cos rt \, dt \tag{11.167}$$

Using a trigonometric identity

$$a_r \frac{1}{\pi} \int_0^\pi \left[ \frac{\sin(t + rt) + \sin(t - rt)}{2} \right] dt \tag{11.168}$$

$$a_r = \frac{1}{2\pi} \int_0^\pi \left[ \sin(r+1)t - \sin(r-1)t \right] dt \tag{11.169}$$

$$a_r = \frac{1}{2\pi} \left[ -\frac{\cos(r+1)t}{(r+1)} + \frac{\cos(r-1)t}{(r-1)} \right]_0^\pi \tag{11.170}$$

$$a_r = \frac{1}{2\pi} \left[ -\frac{(\cos rt \cos t - \sin rt \sin t)}{(r+1)} + \frac{(\cos rt \cos t + \sin rt \sin t)}{(r-1)} \right]_0^\pi \tag{11.171}$$

$$a_r = \frac{1}{2\pi(r^2 - 1)} \left[ -(r-1)(\cos rt \cos t - \sin rt \sin t) + (r+1)(\cos rt \cos t + \sin rt \sin t) \right]_0^\pi$$

$$a_r - \frac{1}{2\pi(r^2 - 1)} [-r \cos rt \cos t + r \sin rt \sin t + \cos rt \cos t \tag{11.172}$$

$$+ \sin rt \sin t + r \cos rt \cos t + r \sin rt \sin t + \cos rt \cos t + \sin rt \sin t]_0^\pi s \tag{11.173}$$

$$a_r = \frac{1}{2\pi(r^2 - 1)} [2r \sin rt \sin t + 2 \cos rt \cos t]_0^\pi \tag{11.174}$$

As the sinusoidal terms vanish when $t = 0$ or $t = \pi$,

$$a_r = \frac{1}{\pi(r^2 - 1)} [\cos rt \cos t]_0^\pi \tag{11.175}$$

$$a_r = \frac{1}{\pi(r^2 - 1)} [\cos r\pi \cos \pi - \cos r0 \cos 0] \tag{11.176}$$

For $r = 1, 3, 5, \ldots \cos r\pi \cos \pi$ is 1 and $\cos r0 \cos 0$ is also 1. The coefficients are therefore all zero.

For $r = 2, 4, 6, \ldots \cos r\pi \cos \pi$ is $-1$; hence,

$$a_r = \frac{-2}{\pi(r^2 - 1)} \qquad (11.177)$$

$$for \; r = 2, 4, 6, \ldots \qquad (11.178)$$

To confirm that first component, $r = 1$, is zero

$$a_1 = \frac{1}{\pi} \int_0^\pi \sin t \cos t \; dt \qquad (11.179)$$

$$a_1 = \frac{1}{\pi} \int_0^\pi \frac{1}{2} \sin 2t \; dt \qquad (11.180)$$

$$a_1 = \frac{1}{2\pi} \left[ \frac{-\cos 2t}{2} \right]_0^\pi = 0 \qquad (11.181)$$

For the $b_r$ coefficients

$$b_r = \frac{1}{\pi} \int_0^\pi \sin t \sin rt \; dt \qquad (11.182)$$

$$br = \frac{1}{\pi} \int_0^\pi \left[ \frac{\cos(t + rt) - \cos(t + rt)}{2} \right] dt \qquad (11.183)$$

$$b_r = \frac{1}{2\pi} \left[ \frac{\sin(rt - t)}{(r + 1)} - \frac{\sin(rt + t)}{(r - 1)} \right]_0^\pi \qquad (11.184)$$

$$b_r = \frac{1}{2\pi} \left[ \frac{(\sin rt \cos t - \cos rt \sin t)}{(r - 1)} - \frac{(\sin rt \cos t + \cos rt \sin t)}{(r + 1)} \right]_0^\pi \qquad (11.185)$$

$$b_r = \frac{1}{2\pi(r^2 - 1)} \left[ (r + 1)(\sin rt \cos t - \cos rt \sin t) - (r + 1)(\sin rt \cos t + \cos rt \sin t) \right]_0^\pi \qquad (11.186)$$

$$b_r = \frac{1}{2\pi(r^2 - 1)} \left[ r \sin rt \cos t - r \cos rt \sin t + \sin rt \cos t - \cos rt \sin t \right.$$
$$\left. - r \sin rt \cos t - r \cos rt \sin t + \sin rt \cos t + \cos rt \sin t \right]_0^\pi \qquad (11.187)$$

$$b_r = \frac{1}{2\pi(r^2 - 1)} \left[ -2r \cos rt \sin t + 2 \sin rt \cos t \right]_0^\pi \qquad (11.188)$$

$$b_r = \frac{1}{\pi(r^2 - 1)} \left[ \sin(r-1)t \right]_0^\pi \qquad (11.189)$$

All the terms for $b_r$ are 0 because sin $(r - 1)\pi$ and sin 0 are 0. However, the first component is not zero.

$$b_1 = \frac{1}{\pi} \int_0^\pi \sin t \sin t \, dt \qquad (11.190)$$

$$b_1 = \frac{1}{\pi} \int_0^\pi \frac{1}{2}(1 - \cos 2t) \, dt \qquad (11.191)$$

$$b_1 = \left[ t - \frac{\sin 2t}{2} \right]_0^\pi = \frac{1}{2\pi}[\pi] = \frac{1}{2} \qquad (11.192)$$

Therefore, the Fourier series of a half-wave rectified sinusoidal wave is

$$v(t) = \frac{1}{\pi} + \frac{1}{2}\sin t + \sum_{r=2}^{\infty} \frac{-2}{\pi(r^2 - 1)} \cos rt \qquad (11.193)$$

$$\text{for } r = 2, 4, 6, \ldots \qquad (11.194)$$

The magnitude ratio, $F$, of the fundamental to the harmonic number, $r$, is the same expression as that for the cosine waveform.

## Example 11.14

Determine the Fourier series expansion for the full-wave rectified voltage, $v$, shown in Figure 11.16.

### Solution

The rectified waveform is a sinusoidal function and therefore is an even function centered about the time $t = 0$.

$$f(t) = -\sin t \quad \text{for} -\pi < t \le 0 \qquad (11.195)$$

$$f(t) = \sin t \quad \text{for } 0 \le t < \pi \qquad (11.196)$$

The coefficients, $b_r$, are all zero and only the cosine amplitudes, $a_r$, are calculated.

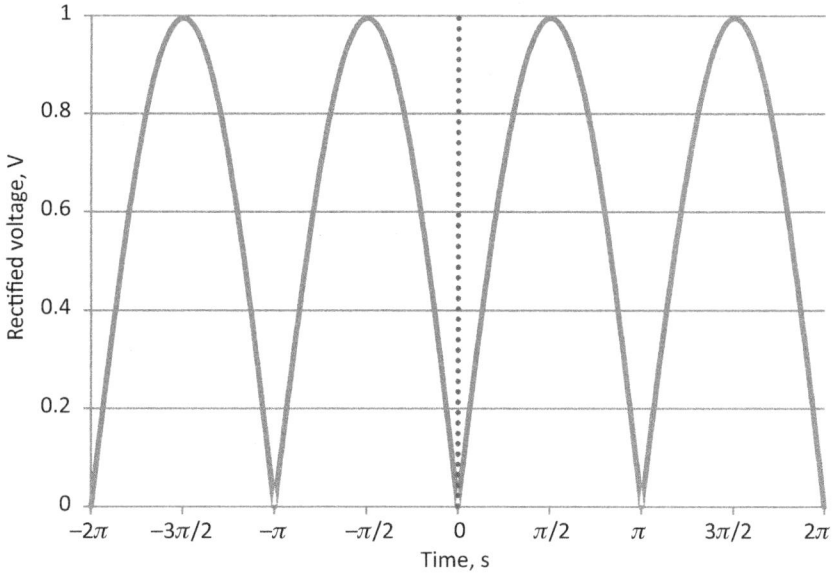

**Figure 11.16** Full rectified voltage, *v* (sine).

The dc term, $a_0$, is

$$a_0 \frac{1}{\pi} \int_{-\pi}^{\pi} f(t) \, dt \qquad (11.197)$$

The integral of the rectified sinusoidal function is

$$\int_{0}^{\pi} |\sin t| \, dt = \int_{-\pi}^{0} |\sin t| \, dt \qquad (11.198)$$

Hence,

$$a_0 \frac{2}{\pi} \int_{0}^{\pi} \sin t \, dt \qquad (11.199)$$

$$a_0 = \frac{2}{\pi} \left[ -\cos t \right]_{0}^{\pi} = \frac{4}{\pi} \qquad (11.200)$$

For the other components

$$a_r = \frac{1}{\pi} \int_{-\pi}^{\pi} |\sin t| \cos rt \, dt \qquad (11.201)$$

As the rectified sine wave is an even function, then

$$a_r = \frac{2}{\pi} \int_0^\pi \sin t \cos rt \, dt \qquad (11.202)$$

These coefficients are twice that of the half sinusoidal wave. Therefore, the Fourier series of a full-wave rectified sine wave is

$$v(t) = \frac{2}{\pi} + \sum_{r=2}^{\infty} \frac{-4}{\pi(r^2 - 1)} \cos rt \qquad (11.203)$$

for

$$r = 2,4,6,... \qquad (11.204)$$

The magnitude ratio, $F$, of the fundamental to the harmonic number, $r$, is the same expression as that for the cosine waveform.

## Comment on the Fourier Series Examples

The ratio of the fundamental component to an individual component is the same whether a sine or cosine is analyzed. However, while the expressions for the Fourier series are similar, the analysis is not as obvious as it first appears. The components are summarized in Table 11.6.

**Table 11.6**
Summary of the Fourier Series Components for Rectified Waveforms

| Waveform | Components | | |
|---|---|---|---|
| | dc ($=2a_0$) | For $r = 1$ | For $r = 2, 4, 6, \ldots$ |
| Cosine half-wave | $\dfrac{1}{\pi}$ | $\dfrac{1}{2}\cos t$ | $\displaystyle\sum_{r=2}^{\infty} \dfrac{-2}{\pi(r^2-1)}(-1)^{r/2}\cos rt$ |
| Cosine full-wave | $\dfrac{2}{\pi}$ | $0$ | $\displaystyle\sum_{r=2}^{\infty} \dfrac{-4}{\pi(r^2-1)}(-1)^{r/2}\cos rt$ |
| Sine half-wave | $\dfrac{1}{\pi}$ | $\dfrac{1}{2}\sin t$ | $\displaystyle\sum_{r=2}^{\infty} \dfrac{-2}{\pi(r^2-1)}\cos rt$ |
| Sine full-wave | $\dfrac{2}{\pi}$ | $0$ | $\displaystyle\sum_{r-2}^{\infty} \dfrac{-4}{\pi(r^2-1)}\cos rt$ |

# List of Symbols

$\alpha$     delay angle (radian)

$\beta$     transistor current gain and extinction angle (°)

$\gamma$     proportion of zero current in a cycle

$\delta$     duty cycle

$\zeta$     damping coefficient

$\eta$     efficiency (%)

$\theta$     angle (radian)

$\omega$     frequency (radian $s^{-1}$)

$\omega_1$     fundamental frequency (radian $s^{-1}$)

$\Delta I$     current interval (current difference) (A)

$\Delta t$     time interval (time difference) (second)

$\Delta T$     temperature interval (temperature difference) (K)

$\Delta V$     voltage interval (voltage difference) (V)

$a_o$     average

$a_r$     amplitude of $r$th cosine Fourier coefficient

$A$     area, voltage-time ($Vs^{-1}$)

ac     alternating current (A)

$b_r$     amplitude of $r$th sine Fourier coefficient

$C$     capacitance (F)

$C_{DG}$     drain-gate capacitance (F)

$C_{DS}$     drain-source capacitance (F)

$C_{GS}$     gate-source capacitance (F)

$C_j$     thermal inertia of a junction (J/K)

$C_s$     thermal inertia of a heat sink (J/K)

| | |
|---|---|
| dc | direct current (A) |
| $d_1$ | proportion of a cycle for a positive interlock delay |
| $d_2$ | proportion of a cycle for a negative interlock delay |
| $E$ | energy (J) |
| $E_1$ | energy (J) |
| $E_2$ | energy (J) |
| $f$ | frequency (Hz) |
| $f_1$ | signal 1 |
| $f_2$ | signal 2 |
| $f_3$ | signal 3 |
| $f_4$ | signal 4 |
| $f_5$ | signal 5 |
| $f_6$ | signal 6 |
| $f_7$ | signal 7 |
| $f(x)$ | function of $x$ |
| $F$ | magnitude |
| $F_f$ | form factor |
| $f_{mean}$ | mean of a function, $f$ |
| $f_{RMS}$ | root mean square of a function, $f$ |
| $g$ | gate signal |
| $i$ | instantaneous current (A) |
| $i_a$ | $a$ phase current (A) |
| $i_b$ | $b$ phase current (A) |
| $i_B$ | base current (A) |
| $i_c$ | $c$ phase current (A) |
| $i_C$ | collector current (A) |
| $i_{cap}$ | capacitor current (A) |
| $i_D$ | diode current and drain current (A) |
| $i_E$ | emitter current (A) |
| $i_G$ | gate current (A) |
| $i_o$ | initial current (A) |
| $I_1$ | current in the cycle (highest) (A) |
| $I_2$ | current at the start of a cycle (lowest) (A) |
| $I_a$ | armature current (A) |
| $I_{av}$ | average current (A) |
| $I_d$ | direct current (A) |
| $I_k$ | constant current (A) |
| $I_L$ | inductor current (A) |
| $I_{Lb}$ | average inductor current at the boundary between continuous and discontinuous current (A) |
| $I_m$ | peak current (A) |
| $I_n$ | amplitude of $n$th harmonic |
| $I_{ob}$ | average output current at the boundary between continuous and discontinuous current (A) |

$I_{RMS}$    root mean square current (A)

$I_0$    dark saturation current (A)

$k$    Boltzmann constant, $1.38 \times 10^{-23}$ (J/K)

$k_m$    dc motor back emf constant (V rad$^{-1}$)

$k_1$    constant (V)

$k_2$    constant (V/nC)

$k_3$    constant (A)

$L$    inductance (H)

$L_a$    armature inductance (H)

$L_s$    supply leakage inductance (H)

$L_{sA}$    supply leakage inductance for converter A (H)

$L_{sB}$    supply leakage inductance for converter B (H)

$n$    parameter

$N_1$    device associated with negative polarity

$N_2$    device associated with negative polarity

$N_3$    device associated with negative polarity .

$N_4$    device associated with negative polarity

$p$    instantaneous power (W) or pulse number of a phase controlled converter

$P$    power (W)

$P_{av}$    average power (W)

$P_o$    base power (W)

$P_1$    device associated with positive polarity

$P_2$    device associated with positive polarity

$P_3$    device associated with positive polarity

$P_4$    device associated with positive polarity

$q$    electron charge, $1.6 \times 10^{-19}$ (C)

$q_G$    gate charge (C)

$R$    resistance ($\Omega$)

$R_a$    armature resistance ($\Omega$)

$R_{DSon}$    drain-source resistance during conduction ($\Omega$)

$R_f$    resistance of a diode in forward conduction ($\Omega$)

$R_p$    resistance $V_d$ divided by $i$ ($\Omega$)

$R_\theta$    thermal resistance (K/W)

$R_{\theta(j-c)}$    internal resistance junction, case (K/W)

$R_{\theta(c-s)}$    resistance case, heat sink (K/W)

$R_{\theta(j-s)}$    resistance junction, heat sink (K/W)

$R_{\theta(s-a)}$    resistance heat sink, ambient (K/W)

$s$    signal and Laplace transform variable

$t$    time (second)

$t_1$    on time (second)

$t_2$    off time (second)

$T$    absolute junction temperature (K)

$T_a$     ambient temperature (K)

$T_c$     case temperature (K)

$T_j$     junction temperature (K)

$T_p$     period (second)

$T_q$     torque (Nm)

$T_1$     first root of the characteristic equation

$T_2$     second root of the characteristic equation

$T_\omega$     time constant (second$^{-1}$)

$v$       instantaneous voltage (V)

$v_{ab}$    line voltage (V)

$v_{an}$    phase voltage (V)

$v_{bc}$    line voltage (V)

$v_{BE}$    base-emitter voltage (V)

$v_{bn}$    phase voltage (V)

$v_c$     cathode voltage (V)

$v_{ca}$    line voltage (V)

$v_{CB}$    collector-base voltage (V)

$v_{CE}$    collector-emitter voltage (V)

$v_{cn}$    phase voltage (V)

$v_{DG}$    drain-gate voltage (V)

$v_{DS}$    drain-source voltage (V)

$v_{GS}$    gate-source voltage (V)

$v_p$     primary voltage (V)

$v_r$     red phase voltage (V)

$v_s$     secondary voltage (V)

$v_y$     yellow phase voltage (V)

$v_o$     output voltage (V)

$V$       supply voltage (V)

$V_a$     armature voltage (V)

$V_{av}$    average voltage (V)

$V_{cap}$    capacitor voltage (V)

$V_d$     dc voltage (V)

$V_{d1}$    dc forward supply voltage (V)

$V_{d2}$    dc reverse supply voltage (V)

$V_{dA}$    dc voltage for converter A (V)

$V_{dB}$    dc voltage for converter B (V)

$V_f$     approximate forward voltage of a diode (V)

$V_m$     peak voltage (V)

$V_o$     voltage difference from initial value or initial value (V)

$V_{RMS}$    root mean square voltage (V)

$V_T$     diode parameter (V)

$Z_{\theta(j-s)}$    transient thermal resistance junction, heat sink (K/W)

# About the Author

**Paul H. Chappell** is an associate professor of Electronics and Computer Science at the University of Southampton in the Faculty of Physical Sciences and Engineering. He graduated from the University of Sussex with a first-class honors degree in electronics and was awarded a Ph.D. in control from the University of Southampton. Dr. Chappell has published over 160 papers. He has extensive teaching experience in the disciplines of electronic and electrical engineering. He teaches postgraduates and undergraduates, covering the subjects of power electronics, electromechanical design, and medical electronics. He is a fellow of the Institution of Engineering Technology, a fellow of the Institute of Physics and Engineering in Medicine, a senior member of the Institute of Electrical and Electronics Engineers, and a member of the Institute of Physics.

# Index

Alternating current (AC), 14
Alternating current (AC) transient suppression, 150

Boost (step-up) converter, 112, 120–28, 166–67
Buck (step-down) converter, 112–20, 165–66

Chaotic behavior, 153–56
Circuit protection, 145–47
Cycloconverter
single-phase, 87–90, 164–65
three-phase, 91–92

DC chopper circuit power, 160–61
DC-to-DC converter
efficiency, 150–53
inductive load, 99–103
motor control, 103–10
regeneration, 110–11
step-down (buck), 112–20, 165–66
step-up (boost), 112, 120–28, 166–67
Defibrillator, 130–34
Digital storage oscilloscope (DSO), 141–44
Diode
free-wheeling (fly-wheel), 26–27, 47, 104, 109
ideal characteristic, 21
power dissipation, 25
Schottky, 28
simple model, 21–24
Zener, 27–28
Direct current (DC), 14. *See also* DC-to-DC converter

Efficiency, DC converter, 150–53
Electric drive, 16
Equivalent circuit, 60–61

Fly-wheel (free-wheeling) diode, 26–27, 47, 104, 109
Forward-biased flow, 21–24
Fourier series
full-wave rectified voltage, 177–84, 188–90
half-wave rectified voltage 171–77, 185–88, 190
overview, 147–150
Free-wheeling (fly-wheel) diode, 26–27, 47, 104, 109
Full-wave rectified voltage, 158–60, 177–84, 188–90

Gate circuit, MOSFET, 56, 167–69
Gate isolation circuit, 31–32
Gate turn-off thyristor (GTO), 42–44

Half-wave rectified voltage, 157–60, 171–77, 185–88, 190
Hall effect current transducer, 143–44
H-bridge circuit, 105–7
Heating and cooling
equivalent circuit, 60–61
thermal resistance, 59–60
transient thermal resistance, 61–63

Ideal power device, 17–18
Inductance load, 36–42
Infinite resistance, 21

Information sources, 19
Insulated gate bipolar transistor (IGBT),
    56–58
Inversion phase-controlled thyristor convertor,
    81–82
Inverter
    single-phase, 93
    three-phase, 93–98

Junction temperature, 59

Logic switching, 129–30
Lowpass filter, 144–45

Maximum critical rate of rise, 32
Metal oxide semiconductor field effect transistor
    (MOSFET)
    circuit protection, 147
    gate characteristic, 167–69
    n-channel circuit, 52–56, 169–71
    overview, 52–56
    p-channel, 52–56
MOSFET. *See* Metal oxide semiconductor
    field effect transistor

Natural commutation, 33, 92
N-channel MOSFET device, 52–56, 169–71

Overlap phase-controlled thyristor converter,
    77–81

Passive component, 16–17
P-channel IGBT device, 56–58
P-channel MOSFET device, 52–56
Phase-controlled thyristor converter
    overlap, 77–81
    overview, 65–67
    single-phase, 65–67
    three-phase, 70–77
Power conversion, 14–16
Power dissipation, 25
Power electronics, 13–19
Practical device, 18–19

Quasi-square waveform, 71

Rectification, 21
    single-phase converter, 67–70
    three-phase converter, 70–77
Regeneration, 110–11
Resistive load, 33–36
Reverse-biased flow, 21–24

Safe operating area, 45–46
Schottky diode, 28
Servomotor transistor, 105
Simple diode model, 21–24
Single-phase cycloconverter, 87–90, 164–65
Single-phase inverter, 93
Single-phase supply inductance load, 36–42
Single-phase supply resistive load, 33–36
Single-phase thyristor converter, 65–67
Six-step waveform, 71
Snubber
    lowpass filter, 145
    thyristor, 32–33
    transistor, 47–52
Step-down (buck) converter, 112–20, 165–66
Step-up (boost) converter, 112, 120–28,
    166–67
Synchronization to AC supply, 135–41

Thermal resistance, 59–63
Three-phase cycloconverter, 91–92
Three-phase thyristor converter, 70–77
Three-phase inverter, 93–98
Thyristor. *See also* Thyristor converter,
    phase-controlled
    gate isolation circuit, 31–32
    gate turn-off, 42–44
    inductance load, 36–42
    natural commutation, 33
    overview, 29–31
    resistive load, 33–36
    single-phase supply, 33–42
    snubber, 32–33
    triac, 42
Thyristor converter, phase-controlled
    AC motor, 65–67
    DC motor, 84–86
    inversion, 81–82
    overlap, 77–81
    overview, 65–67
    power systems, 82–84
    single-phase, 67–70
    three-phase, 70–77
Total harmonic distortion (THD), 149–50
Transcranial magnetic stimulator (TMS),
    134
Transient suppression, 150
Transient thermal resistance, 61–63

Transistor
  IGBT, 56–58
  MOSFET, 52–56
  overview, 45
  power level and frequency, 58
  safe operating areas, 45–46
  snubber, 47–52
Triac, 42

Voltage, transmission circuit, 162–64

Waveform parameters, 17

Zener diode, 27–28
Zero leakage current, 18
Zero resistance, 21
Zero-crossing-point, 135

# Recent Artech House Titles
# in Power Engineering

Dr. Jianhui Wang, Series Editor

*The Advanced Smart Grid: Edge Power Driving Sustainability,*
Andres Carvallo and John Cooper

*Battery Management Systems for Large Lithium Ion Battery Packs,*
Davide Andrea

*Battery Power Management for Portable Devices,*
Yevgen Barsukov and Jinrong Qian

*Designing Control Loops for Linear and Switching Power Supplies:
A Tutorial Guide,* Christophe Basso

*Electric Systems Operations: Evolving to the Modern Grid,*
Mani Vadari

*Energy Harvesting for Autonomous Systems,* Stephen Beeby and
Neil White

*GIS for Enhanced Electric Utility Performance,* Bill Meehan

*Introduction to Power Electronics,* Paul H. Chappell

*Power Line Communications in Practice,* Xavier Carcelle

*Power System State Estimation,* Mukhtar Ahmad

*A Systems Approach to Lithium-Ion Battery Management,*
Phil Weicker

*Synergies for Sustainable Energy,* Elvin Yüzügüllü

For further information on these and other Artech House
titles,including previously considered out-of-print books now
available through our In-Print-Forever® (IPF®) program, contact:

| Artech House | Artech House |
|---|---|
| 685 Canton Street | 16 Sussex Street |
| Norwood, MA 02062 | London SW1V 4RW UK |
| Phone: 781-769-9750 | Phone: +44 (0)20 7596-8750 |
| Fax: 781-769-6334 | Fax: +44 (0)20 7630-0166 |
| e-mail: artech@artechhouse.com | e-mail: artech-uk@artechhouse.com |

Find us on the World Wide Web at: www.artechhouse.com